REVOLUTIONARY ENOLOGY

THE NEW ERA OF ENOLOGY AND VITICULTURE

DAVID SANDUA

Revolutionary enology.
© David Sandua 2024. All rights reserved.
eBook & Paperback Edition.

*"Winemaking is a balance between tradition and innovation.
We should not be afraid to experiment."*

Christian Moueix

INDEX

I. INTRODUCTION 12
    Overview of traditional oenology and viticulture practices 13
    The impetus for innovation in the wine industry 14
    Defining the new era of revolutionary oenology 15

II. HISTORICAL CONTEXT OF OENOLOGY 16
    Ancient winemaking techniques and their evolution 17
    The role of the Industrial Revolution in viticulture 18
    Twentieth-century advancements and their limitations 19

III. CLIMATE CHANGE AND VITICULTURE 20
    Impact of global warming on grape cultivation 21
    Adaptation strategies for vineyards 22
    Predictive models for future viticulture practices 23

IV. PRECISION VITICULTURE 24
    Introduction to precision agriculture in vineyards 25
    Technological tools for monitoring and decision-making 26
    Case studies of successful precision viticulture implementation 27

V. GENETIC INNOVATION IN GRAPE CULTIVATION 28
    Genetically modified vines for disease resistance 29
    Clonal selection and its benefits for quality and yield 30
    Ethical considerations and consumer acceptance 31

VI. SOIL MANAGEMENT AND TERROIR 32
    Advances in soil analysis techniques 33
    The relationship between soil health and wine quality 34
    Innovative practices for sustainable soil management 35

VII. WATER MANAGEMENT IN VINEYARDS 36
    New irrigation technologies and strategies 37
    Water conservation and its importance in viticulture 38
    The effects of water stress on grape quality 39

VIII. SUSTAINABLE VITICULTURE 40
    Principles of sustainability in wine production 41
    Organic and biodynamic farming practices 42
    Certification programs and consumer awareness 43

IX. BIODIVERSITY IN VINEYARDS 44
    The role of biodiversity in ecosystem health 45
    Strategies for promoting biodiversity 46
    Benefits of biodiversity for disease control and yield 47

X. ADVANCES IN FERMENTATION TECHNOLOGY 48
    Innovations in yeast and bacteria strains 49
    Controlled fermentation environments 50
    The impact of fermentation technology on wine flavor profiles 51

XI. WINE AGING AND MATURATION 52
    New approaches to barrel aging 53

- Alternative aging materials and methods 54
- The science behind wine aging and its effects on quality 55

XII. WINE CHEMISTRY AND ANALYTICS 56
- The role of chemistry in understanding wine 57
- Advanced analytical techniques for wine composition 58
- Applications of chemometrics in oenology 59

XIII. SENSORY SCIENCE AND CONSUMER PREFERENCES 60
- The study of sensory perception in wine tasting 61
- The influence of sensory attributes on consumer choices 62
- The development of sensory profiles for wine branding 63

XIV. WINE PACKAGING INNOVATIONS 64
- Eco-friendly and sustainable packaging solutions 65
- The evolution of wine bottle design and materials 66
- The impact of packaging on wine preservation and sales 67

XV. WINE MARKETING AND DIGITAL TRANSFORMATION 68
- The role of digital marketing in the wine industry 69
- E-commerce and direct-to-consumer sales strategies 70
- Leveraging social media and technology for brand building 71

XVI. GLOBAL WINE TRADE AND ECONOMICS 72
- The current state of the global wine market 73
- Trade agreements and their impact on viticulture 74
- Economic challenges and opportunities in the wine industry 75

XVII. WINE EDUCATION AND SOMMELIER SCIENCE 76
- The evolution of wine education programs 77
- The science behind sommelier training 78
- The future of wine expertise and certification 79

XVIII. WINE TOURISM AND EXPERIENCE ECONOMY 80
- The growth of wine tourism 81
- Creating immersive experiences for wine enthusiasts 82
- The economic impact of wine tourism on regions 83

XIX. REGULATORY CHANGES AND WINE LAW 84
- Recent changes in wine legislation 85
- The impact of regulations on production and distribution 86
- International wine laws and their harmonization 87

XX. WINE AND HEALTH 88
- Research on wine's health benefits and risks 89
- The role of moderation in wine consumption 90
- Wine in the context of dietary guidelines 91

XXI. THE ROLE OF WOMEN IN OENOLOGY 92
- Historical contributions of women in winemaking 93
- Current trends in female leadership in the wine industry 94
- The future of gender diversity in oenology and viticulture 95

XXII. EMERGING WINE REGIONS 96
- Characteristics of new and promising wine regions 97
- Challenges faced by emerging wine producers 98
- The potential impact of new regions on the global wine scene 99

XXIII. URBAN WINERIES AND MICRO-VITICULTURE 100

The rise of urban winemaking 101
Techniques and challenges of micro-viticulture 102
The role of urban wineries in local economies 103

XXIV. WINE AND TECHNOLOGY: THE ROLE OF AI 104
Artificial intelligence in wine pattern recognition 105
AI-driven decision-making in viticulture 106
The future of AI in personalized wine recommendations 107

XXV. ROBOTICS AND AUTOMATION IN VITICULTURE 108
The use of robots for vineyard tasks 109
Benefits and limitations of automation 110
Case studies of robotics in action 111

XXVI. WINE AND THE CIRCULAR ECONOMY 112
Principles of the circular economy applied to winemaking 113
Upcycling and waste reduction in the wine industry 114
Case examples of circular economy practices 115

XXVII. THE IMPACT OF SOCIAL MOVEMENTS ON WINE PRODUCTION 116
Fair trade and ethical sourcing in viticulture 117
The influence of environmental movements on winemaking practices 118
Social responsibility and corporate governance in the wine industry 119

XXVIII. THE SCIENCE OF WINE FLAVORS AND AROMAS 120
Understanding the molecular composition of wine flavors 121
The role of terpenes and esters in aroma profiles 122
Advances in flavor enhancement and manipulation 123

XXIX. THE FUTURE OF WINE CRITICISM AND REVIEWS 124
The changing landscape of wine journalism 125
The role of online platforms and influencers 126
The impact of technology on wine rating systems 127

XXX. WINE AND CLIMATE ADAPTATION STRATEGIES 128
Breeding grapes for climate resilience 129
Vineyard management adjustments for extreme weather 130
Long-term strategies for climate change mitigation 131

XXXI. INNOVATIONS IN GRAPE HARVESTING 132
Mechanized vs. hand-picking: efficiency and quality 133
The development of selective harvesting machinery 134
The impact of harvest timing on wine characteristics 135

XXXII. THE INTERSECTION OF ART AND SCIENCE IN WINEMAKING 136
The balance between creativity and technical knowledge 137
The role of the winemaker as an artist 138
Collaborations between scientists and winemakers 139

XXXIII. THE ECONOMICS OF WINE PRODUCTION 140
Cost analysis of traditional vs. innovative practices 141
Investment in technology and its ROI for wineries 142
Economic models for sustainable viticulture 143

XXXIV. THE ROLE OF CONSORTIA AND WINE ASSOCIATIONS 144
The function of wine consortia in regulation and promotion 145
Collaborative efforts for research and development 146
Case studies of successful consortia initiatives 147

XXXV. WINE LABELING AND CONSUMER INFORMATION 148
- THE EVOLUTION OF WINE LABELS AS A SOURCE OF INFORMATION 149
- THE ROLE OF LABELS IN COMMUNICATING WINE QUALITY AND ORIGIN 150
- REGULATORY CHANGES AND TRENDS IN WINE LABELING 151

XXXVI. THE INFLUENCE OF WINE ON CULTURE AND SOCIETY 152
- WINE AS A CULTURAL SYMBOL AND ITS HISTORICAL SIGNIFICANCE 153
- THE SOCIAL DYNAMICS OF WINE CONSUMPTION 154
- WINE'S ROLE IN CONTEMPORARY CULTURAL PRACTICES 155

XXXVII. THE PSYCHOLOGY OF WINE TASTING 156
- COGNITIVE ASPECTS OF TASTE AND AROMA PERCEPTION 157
- THE INFLUENCE OF CONTEXT AND EXPECTATIONS ON WINE TASTING 158
- EXPERIMENTAL STUDIES ON THE PSYCHOLOGY OF WINE PREFERENCES 159

XXXVIII. THE INTEGRATION OF WINE IN CULINARY ARTS 160
- WINE PAIRING PRINCIPLES AND CONTEMPORARY TRENDS 161
- THE COLLABORATION BETWEEN CHEFS AND WINEMAKERS 162
- THE ROLE OF WINE IN GASTRONOMY AND FINE DINING 163

XXXIX. WINE AND THE GIG ECONOMY 164
- THE RISE OF FREELANCE WORK IN THE WINE INDUSTRY 165
- THE IMPACT OF THE GIG ECONOMY ON TRADITIONAL WINE JOBS 166
- OPPORTUNITIES AND CHALLENGES FOR GIG WORKERS IN OENOLOGY 167

XL. THE ROLE OF WINE IN SUSTAINABLE DEVELOPMENT GOALS 168
- ALIGNING WINE PRODUCTION WITH THE UN'S SDGs 169
- CASE STUDIES OF WINERIES CONTRIBUTING TO SUSTAINABLE DEVELOPMENT 170
- THE POTENTIAL FOR WINE TO DRIVE SOCIAL AND ENVIRONMENTAL PROGRESS 171

XLI. THE IMPACT OF EXCHANGE RATES AND TARIFFS 172
- THE INFLUENCE OF CURRENCY FLUCTUATIONS ON WINE TRADE 173
- THE EFFECTS OF TARIFFS ON INTERNATIONAL WINE MARKETS 174
- STRATEGIES FOR WINERIES TO MITIGATE ECONOMIC RISKS 175

XLII. THE ROLE OF WINE IN INTERNATIONAL DIPLOMACY 176
- WINE AS A TOOL FOR CULTURAL EXCHANGE AND DIPLOMACY 177
- HISTORICAL INSTANCES OF WINE INFLUENCING DIPLOMATIC RELATIONS 178
- THE POTENTIAL FOR WINE TO BRIDGE INTERNATIONAL DIVIDES 179

XLIII. WINE AUTHENTICATION AND FRAUD PREVENTION 180
- TECHNOLOGIES FOR ENSURING WINE AUTHENTICITY 181
- THE GLOBAL IMPACT OF WINE FRAUD 182
- LEGAL AND REGULATORY MEASURES TO COMBAT COUNTERFEIT WINES 183

XLIV. THE SCIENCE OF WINE PRESERVATION 184
- ADVANCES IN WINE STORAGE TECHNOLOGY 185
- THE CHEMISTRY OF WINE SPOILAGE AND PRESERVATION 186
- INNOVATIONS IN EXTENDING WINE SHELF LIFE 187

XLV. THE ROLE OF WINE IN FESTIVALS AND EVENTS 188
- THE ECONOMIC IMPACT OF WINE FESTIVALS 189
- WINE'S CONTRIBUTION TO EVENT EXPERIENCES 190
- CASE STUDIES OF SUCCESSFUL WINE-CENTERED EVENTS 191

XLVI. THE INTERSECTION OF WINE AND TECHNOLOGY STARTUPS 192
- THE EMERGENCE OF WINE TECH STARTUPS 193
- DISRUPTIVE TECHNOLOGIES RESHAPING THE WINE INDUSTRY 194
- THE CHALLENGES AND SUCCESSES OF WINE TECH ENTREPRENEURSHIP 195

XLVII. WINE EDUCATION AND TECHNOLOGY TRANSFER 196
THE IMPORTANCE OF KNOWLEDGE DISSEMINATION IN VITICULTURE 197
PLATFORMS AND METHODS FOR TECHNOLOGY TRANSFER 198
THE ROLE OF ACADEMIC INSTITUTIONS IN WINE EDUCATION 199

XLVIII. THE FUTURE OF WINE DISTRIBUTION CHANNELS 200
TRENDS IN WINE RETAIL AND DISTRIBUTION 201
THE IMPACT OF DIRECT SHIPPING AND ONLINE SALES 202
THE FUTURE LANDSCAPE OF WINE DISTRIBUTION 203

XLIX. THE ROLE OF WINE IN COMMUNITY BUILDING 204
WINE CLUBS AND THEIR CONTRIBUTION TO SOCIAL COHESION 205
THE IMPACT OF LOCAL WINERIES ON COMMUNITY DEVELOPMENT 206
WINE AS A MEDIUM FOR COMMUNITY ENGAGEMENT AND SUPPORT 207

L. THE ETHICS OF WINE PRODUCTION AND CONSUMPTION 208
ETHICAL CONSIDERATIONS IN LABOR PRACTICES AND SOURCING 209
THE SOCIAL RESPONSIBILITIES OF WINE CONSUMERS 210
THE ROLE OF ETHICS IN SHAPING THE FUTURE OF WINE 211

LI. CONCLUSION 212
SUMMARY OF THE REVOLUTIONARY CHANGES IN OENOLOGY AND VITICULTURE 213
REFLECTION ON THE IMPLICATIONS FOR THE WINE INDUSTRY'S FUTURE 214
CHALLENGES AND OPPORTUNITIES FOR ONGOING RESEARCH AND INNOVATION 215

BIBLIOGRAPHY 216

I. INTRODUCTION

The field of oenology and viticulture has reached a critical juncture, with the need for innovation becoming ever more pressing. This marks the beginning of a revolutionary era in wine production, where traditional methods are being redefined and new approaches are being embraced. As the demands of consumers evolve and climate change poses challenges to grape cultivation, the time is ripe for a transformation in the industry. Advances in technology, such as precision viticulture and the use of AI and drones in vineyard management, are pushing boundaries and opening new possibilities for winemakers. Additionally, the integration of sustainability practices and organic viticulture are shaping a more responsible and environmentally-friendly future for the wine industry. This introduction sets the stage for a deeper exploration of the groundbreaking developments that are reshaping oenology and viticulture in this new era.

Overview of traditional oenology and viticulture practices

In examining the traditional practices of oenology and viticulture, it is evident that these time-honored techniques have played a crucial role in shaping the wine industry as we know it today. For centuries, oenologists and viticulturists have relied on methods passed down through generations, involving manual labor in the vineyards, natural fermentation processes, and minimal intervention in winemaking. These practices, rooted in history and tradition, have laid the foundation for the production of high-quality wines around the world. While these traditional methods have produced exceptional wines, the onset of a new era in oenology and viticulture beckons the need for innovation and modernization. As advancements in technology and scientific knowledge continue to revolutionize the field, it is imperative to explore how these changes can enhance the quality, sustainability, and efficiency of wine production in the future.

The impetus for innovation in the wine industry

The impetus for innovation in the wine industry stems from a combination of factors that have converged to create a pressing need for change. With shifting consumer preferences, environmental challenges, and technological advancements reshaping the landscape, wine producers are compelled to adapt and evolve to stay competitive in the market. The demand for sustainable practices and organic wines has pushed the industry towards greener viticultural methods, such as biodynamic farming and precision viticulture. Additionally, advancements in sensor technologies, data analytics, and automation have revolutionized winemaking processes, allowing for greater precision, efficiency, and quality control. These changes are not only reshaping the industry but also shaping the future of oenology and viticulture, paving the way for continued innovation and growth in the field. The challenges moving forward will involve striking a balance between tradition and innovation, sustainability and profitability, while meeting the evolving demands of consumers and the environment.

Defining the new era of revolutionary oenology

Defining the new era of revolutionary oenology involves a paradigm shift in the traditional practices of winemaking. This era is characterized by a fusion of creativity, innovation, and science, pushing boundaries to produce exceptional wines. One aspect of this revolution is the integration of precision viticulture, where technologies such as drones, sensors, and GIS mapping are used to optimize vineyard management. These tools provide valuable data on soil health, moisture levels, and grape ripeness, enabling winemakers to make informed decisions to enhance grape quality. Furthermore, advancements in fermentation techniques, including the use of indigenous yeasts and new fermentation vessels, are paving the way for unique flavor profiles and improved complexity in wines. The synergy between tradition and innovation in oenology is reshaping the industry, ushering in a new era of excellence and possibility for winemakers and wine enthusiasts alike.

II. HISTORICAL CONTEXT OF OENOLOGY

The historical context of oenology provides a rich tapestry that has shaped the development of winemaking throughout the ages. From ancient civilizations like the Greeks and Romans who established the foundations of viticulture, to the monks of the Middle Ages who meticulously recorded their winemaking techniques, each era has contributed to the evolution of oenology. The Renaissance period saw a revival of interest in wine production, with advancements in irrigation systems and the classification of grape varieties. The Industrial Revolution brought about technological innovations like the steam-powered press and the wine bottling machine, revolutionizing the scale and efficiency of winemaking. By examining these historical milestones, we can gain insights into how traditional practices have influenced modern oenology, providing a foundation upon which the new era of innovative techniques and methodologies can flourish.

Ancient winemaking techniques and their evolution

Ancient winemaking techniques have served as the foundation for the evolution of oenology throughout history. From the earliest civilizations in Mesopotamia to the sophisticated practices of the Romans, traditional methods have shaped contemporary winemaking processes. These ancient techniques encompassed a deep understanding of grape cultivation, fermentation, and preservation methods. Over time, with the advancement of technology and scientific knowledge, these practices have been refined and enhanced to improve quality and consistency. The use of modern equipment, such as stainless steel tanks and temperature-controlled fermentation, has revolutionized winemaking, allowing for greater control over the process. However, despite these advancements, the importance of traditional techniques, such as hand-harvesting and oak barrel aging, is still recognized for their contribution to flavor and complexity in wines. The synthesis of ancient wisdom and modern innovation continues to drive the evolution of winemaking, pushing the boundaries of what is possible in oenology and viticulture.

The role of the Industrial Revolution in viticulture

The Industrial Revolution played a pivotal role in transforming viticulture practices, reshaping the landscape of wine production. With the advent of steam-powered machinery, the labor-intensive processes of planting, pruning, and harvesting grapes became more efficient and cost-effective. The mechanization of winemaking equipment, such as presses and crushers, also led to higher yields and improved quality control. This revolution in technology not only increased productivity but also allowed vineyards to expand their operations and reach new markets. Additionally, the development of new transportation networks facilitated the distribution of wine on a global scale, further propelling the industry forward. As a result, the Industrial Revolution significantly influenced the evolution of viticulture, laying the foundation for the modern oenology practices we see today.

Twentieth-century advancements and their limitations

In the twentieth century, oenology and viticulture experienced significant advancements that revolutionized the industry. Innovations in fermentation techniques, such as the introduction of controlled temperature and yeast strains, resulted in improved quality and consistency of wines. Additionally, the use of modern analytical tools and equipment allowed for precise monitoring of grape maturity and vineyard conditions, leading to more efficient grape growing and harvesting practices. Despite these advancements, limitations exist within the field. For instance, while technology has enhanced production processes, it has also raised concerns about the loss of traditional winemaking practices and the potential homogenization of wine styles. Furthermore, environmental sustainability remains a pressing issue, as the intensification of viticultural practices may have adverse effects on ecosystems and biodiversity. As we continue to navigate this new era of oenology and viticulture, it is crucial to balance innovation with tradition, and prioritize sustainability to ensure a prosperous future for the industry.

III. CLIMATE CHANGE AND VITICULTURE

Climate change poses significant challenges for the viticulture industry, impacting grape growing regions worldwide. As temperatures rise, shifts in precipitation patterns and increased frequency of extreme weather events have the potential to disrupt traditional grape growing practices. Changes in temperature can affect grape ripening, altering sugar and acidity levels in the berries, thus influencing wine quality. Additionally, rising temperatures may lead to earlier harvests and changes in grape phenology, impacting grape yields and affecting the overall wine production. Mitigation strategies such as adjusting vineyard management practices, implementing new trellising techniques, and exploring grape varietals better suited to changing climate conditions are crucial for the sustainability of the industry. By addressing the challenges of climate change through innovative approaches, viticulturists can adapt to the evolving environmental conditions and maintain the quality and diversity of wines produced in the new era of oenology.

Impact of global warming on grape cultivation

Global warming is having a significant impact on grape cultivation, posing challenges to the centuries-old traditions of oenology and viticulture. Rising temperatures, changing weather patterns, and increased occurrences of extreme events such as droughts and heatwaves are altering the conditions suitable for grape growth. These changes can affect grape quality, yield, and even the types of grape varietals that can thrive in certain regions. In response to these challenges, researchers and winemakers are exploring innovative solutions such as the development of new grape varieties that are more resistant to heat stress, implementing water-saving irrigation practices, and adopting precision viticulture techniques to optimize grape production. As the effects of global warming intensify, it is crucial for the wine industry to adapt and innovate to ensure the sustainability and longevity of grape cultivation practices.

Adaptation strategies for vineyards

In response to the challenges posed by climate change and evolving environmental conditions, vineyards have increasingly turned to adaptation strategies to safeguard their yield and quality. One notable approach involves the implementation of precision viticulture techniques, utilizing data-driven analysis to optimize vineyard management practices. By employing satellite imagery, soil sensors, and weather monitoring systems, growers can make informed decisions regarding irrigation, fertilization, and pest control, mitigating risks and enhancing sustainability. Furthermore, the adoption of climate-resilient grape varieties and rootstocks has proven effective in ensuring vineyard resilience against extreme weather events and shifting climate patterns. These adaptation strategies not only serve to protect vineyard productivity but also contribute to the preservation of terroir characteristics and the maintenance of grape quality. As vineyards continue to face environmental uncertainties, the integration of innovative practices will be crucial in adapting to and thriving in this new era of oenology and viticulture.

Predictive models for future viticulture practices

One key aspect of the new era of oenology and viticulture lies in the development and implementation of predictive models for future viticulture practices. These models utilize advanced technologies such as artificial intelligence, machine learning, and data analytics to forecast various variables affecting grape-growing conditions, including climate fluctuations, soil composition, and pest outbreaks. By analyzing vast amounts of historical and real-time data, these predictive models can offer invaluable insights into optimal planting strategies, irrigation schedules, harvest timings, and disease management protocols. Furthermore, these models can enhance resource efficiency, sustainability, and overall crop yields, leading to a more environmentally friendly and economically viable viticulture industry. As the field of predictive modeling continues to evolve, researchers and practitioners alike must harness its potential to drive further innovation and advancement in oenology and viticulture for years to come.

IV. PRECISION VITICULTURE

has emerged as a pivotal advancement in modern oenology and viticulture, revolutionizing the way grape cultivation is approached. By utilizing cutting-edge technologies such as drones, satellite imaging, and sensors, precision viticulture allows for a targeted and data-driven approach to vineyard management. This method enables growers to monitor and optimize various aspects of grape production, including irrigation, fertilization, and pest control, with unprecedented accuracy and efficiency. Case studies have demonstrated the tangible benefits of precision viticulture, showcasing increased grape quality, yield, and sustainability. The integration of big data and artificial intelligence further enhances decision-making processes, paving the way for a more sustainable and economically viable future for the wine industry. As precision viticulture continues to evolve, its potential in shaping the future of grape production is undeniable, driving further research and innovation in the field.

Introduction to precision agriculture in vineyards

Precision agriculture in vineyards represents a significant advancement in the field of oenology and viticulture. This strategy involves the use of innovative technologies such as drones, sensors, and GPS to collect accurate data and optimize various aspects of grape production. By precisely monitoring factors like soil moisture, nutrient levels, and plant health, vineyard managers can make informed decisions to enhance grape quality and yield. Case studies have shown that the implementation of precision agriculture practices can lead to improved resource efficiency, reduced environmental impact, and higher economic returns. Through the integration of data-driven decision-making and targeted interventions, precision agriculture holds great potential for transforming traditional vineyard management practices. As we continue to navigate this new era of oenology and viticulture, the adoption of precision agriculture techniques will play a crucial role in shaping the future of grape production and wine quality.

Technological tools for monitoring and decision-making

Technological tools for monitoring and decision-making have revolutionized the field of oenology and viticulture, providing winemakers and vineyard managers with unprecedented levels of data and analysis. From precision viticulture techniques utilizing drones and satellite imagery to sensor technology monitoring soil conditions and climate variables, these tools have allowed for real-time decision-making that can optimize grape quality and yield. For example, the use of automated weather stations and moisture sensors can help predict disease outbreaks or drought conditions, enabling proactive interventions. Furthermore, advancements in data analytics and artificial intelligence have enabled the creation of predictive models for vineyard management, fermentation processes, and even wine quality assessments. By harnessing the power of these technologies, the wine industry can not only improve efficiency and sustainability but also elevate the overall quality of wines produced. The integration of technological tools into traditional practices marks a new era in oenology and viticulture, where innovation and data-driven decisions are key drivers of success.

Case studies of successful precision viticulture implementation

Case studies of successful precision viticulture implementation highlight the transformative potential of innovative technologies and methodologies in the wine industry. For example, the use of drones equipped with multispectral cameras has enabled vineyard managers to identify variability in soil and grape ripeness levels, leading to targeted interventions and improved grape quality. Additionally, sensor networks and data analytics have revolutionized irrigation practices, allowing for precise water management tailored to the specific needs of each vine. These advancements have not only optimized resource utilization but also enhanced overall vineyard productivity and sustainability. By showcasing real-world applications of precision viticulture, these case studies underscore the importance of embracing technological advancements in the pursuit of excellence in oenology and viticulture. As the industry continues to evolve, integrating these solutions into everyday practices will be essential for ensuring continued success and competitiveness in the global market.

V. GENETIC INNOVATION IN GRAPE CULTIVATION

Genetic innovation in grape cultivation has become a key area of interest in the field of oenology and viticulture, offering promising solutions to longstanding challenges. With advancements in biotechnology and genetic engineering, researchers are now able to manipulate the genetic makeup of grapevines to enhance traits such as disease resistance, yield, and flavor profiles. By introducing genetic modifications, scientists can create new grape varieties that are better suited to specific terroirs or climate conditions, ultimately improving the quality and consistency of wines produced. Additionally, genetic innovation allows for the preservation of rare or ancient grape varieties that may be at risk of extinction due to environmental changes or disease outbreaks. As this technology continues to evolve, the potential for genetic innovation in grape cultivation to revolutionize the wine industry is vast, opening up new opportunities for experimentation and creativity in winemaking.

Genetically modified vines for disease resistance

In the realm of viticulture, the emergence of genetically modified vines for disease resistance represents a groundbreaking development in the quest for sustainable grape cultivation practices. Through the integration of biotechnological advancements, researchers have been able to engineer vines with enhanced resistance to common diseases such as powdery mildew and downy mildew. By introducing specific genes into the grapevine genome, these genetically modified vines demonstrate increased tolerance to pathogens, reducing the need for chemical pesticides and fungicides. This not only benefits the environment by decreasing chemical inputs but also improves the overall health and longevity of vineyards. However, concerns regarding the potential impact on biodiversity and consumer perception linger, prompting further research and dialogue in the field of viticulture. As this technology continues to evolve, it holds the potential to revolutionize the way vineyards are managed and sustained in the future.

Clonal selection and its benefits for quality and yield

Clonal selection in viticulture involves the careful selection and propagation of grapevine varieties with desirable traits to optimize quality and yield. By identifying and replicating vines with exceptional characteristics such as disease resistance, flavor profiles, and adaptation to specific terroirs, growers can improve overall grape quality and increase production efficiency. Through scientific research and rigorous testing, certain clones have been proven to outperform their counterparts in terms of consistency and resilience, leading to enhanced wine quality and higher yields. Clonal selection offers a strategic advantage to winemakers seeking to differentiate their products in a competitive market, allowing for the production of wines with distinct flavors and characteristics that resonate with consumers. This innovative approach to vineyard management is revolutionizing the industry, shaping a new era of oenology and viticulture characterized by precision, excellence, and sustainability.

Ethical considerations and consumer acceptance

Ethical considerations play a crucial role in the acceptance of new advancements in oenology and viticulture. Consumer awareness of sustainability, organic practices, and labor conditions has been increasing, leading to a demand for ethical production methods in the wine industry. While innovations such as precision viticulture and biodynamic farming can enhance quality and productivity, they must also align with ethical standards to gain consumer trust. Fair labor practices, environmental stewardship, and transparency in production are essential factors that shape consumer perceptions and purchasing decisions. Therefore, wineries that prioritize ethical considerations stand to gain a competitive edge in the market and foster long-term relationships with consumers. Moving forward, it is imperative for researchers and practitioners in oenology to prioritize ethical principles to ensure the sustainable growth and acceptance of new technologies in the field.

VI. SOIL MANAGEMENT AND TERROIR

Soil management and terroir play a crucial role in shaping the quality and distinct characteristics of wine grapes. The composition of the soil, including its nutrient content, pH levels, water retention capacity, and drainage, can significantly influence grapevine growth and the final flavor profile of the wine produced. By implementing sustainable soil management practices, such as cover cropping, composting, and minimal tillage, viticulturists can improve soil health, increase biodiversity, and enhance the expression of terroir in the wine. Understanding the unique terroir of a vineyard, which encompasses the soil type, climate, topography, and microclimate, allows winemakers to produce wines that truly reflect the essence of a specific region. Through careful soil management practices and a deep appreciation for terroir, oenologists can create wines that are not only of exceptional quality but also truly representative of their origin.

Advances in soil analysis techniques

Advances in soil analysis techniques have been a game-changer in the field of oenology and viticulture, providing researchers and winemakers with a deeper understanding of the complex interaction between soil composition, grapevine health, and wine quality. Traditional methods such as manual soil sampling and analysis have been replaced by cutting-edge technologies like remote sensing, geostatistics, and DNA sequencing, allowing for more precise and comprehensive soil assessments. These advancements have enabled scientists to identify specific nutrient deficiencies, monitor soil moisture levels, and assess the impact of climate change on vineyard soils. For example, the use of high-resolution soil sampling combined with geographic information systems (GIS) has revolutionized precision agriculture, leading to increased vineyard productivity and sustainability. By incorporating these innovative techniques into their practices, oenologists and viticulturists can make informed decisions to optimize grape quality and maximize wine production, ultimately shaping the future of the industry.

The relationship between soil health and wine quality

One key aspect that is often overlooked in the pursuit of high-quality wine production is the relationship between soil health and wine quality. Soil health plays a crucial role in providing the necessary nutrients for grapevines to thrive, impacting the flavor profile and overall quality of the wine produced. Different soil types can influence the characteristics of the grapes, with factors such as drainage, pH levels, and mineral composition all playing a role in grape development. By understanding and optimizing soil health, winemakers can enhance the terroir of their wines and create unique, distinctive flavors that reflect the environment in which the grapes were grown. Research in this area is essential for advancing oenology and viticulture, as it allows for a deeper understanding of how soil health can directly impact wine quality, leading to more innovative and sustainable practices in the industry.

Innovative practices for sustainable soil management

Innovative practices for sustainable soil management are essential in the realm of oenology and viticulture, as the quality and health of the soil directly impact the grapes' growth and ultimately the wine produced. One such practice gaining traction is regenerative agriculture, a holistic approach that aims to restore soil health through carbon sequestration, biodiversity promotion, and reducing the use of agrochemicals. By incorporating cover crops, rotational grazing, and composting, vineyards can improve soil structure, water retention, and nutrient availability, leading to healthier vines and higher quality grapes. Additionally, precision agriculture technologies such as soil sensors and drones can provide real-time data on soil moisture, nutrient levels, and pest infestations, enabling growers to make informed decisions that optimize resource use and minimize environmental impact. These innovative practices not only ensure sustainable soil management but also contribute to the overall resilience and longevity of vineyards in the face of climate change and other challenges.

VII. WATER MANAGEMENT IN VINEYARDS

Water management in vineyards plays a crucial role in ensuring the health and productivity of grapevines. The availability of water and its controlled distribution can significantly impact vine growth, fruit development, and overall grape quality. With the increasing unpredictability of climate patterns and the growing focus on sustainability, vineyard managers are increasingly turning to innovative water management techniques to optimize water use efficiency. From precision irrigation systems to soil moisture sensors, these technologies allow for more precise control over water delivery, reducing waste and promoting healthier vine growth. Case studies have shown that strategic water management practices not only conserve resources but also improve grape quality and yield. As the wine industry continues to evolve, incorporating advanced water management strategies will be essential for vineyards to adapt to changing environmental conditions and ensure long-term sustainability.

New irrigation technologies and strategies

In the realm of oenology and viticulture, the integration of new irrigation technologies and strategies has been identified as a crucial component in advancing the quality and sustainability of grape production. With a growing focus on water conservation and precision agriculture, innovative techniques such as drip irrigation, soil moisture sensors, and deficit irrigation have gained traction in vineyards worldwide. Recent studies have shown that these advancements not only improve water efficiency and reduce environmental impact but also enhance grape quality by controlling vine vigor and influencing fruit development. For instance, the implementation of precision irrigation systems in hilly terrains has demonstrated significant improvements in grape yield and quality, showcasing the potential of these technologies in challenging growing conditions. As the field of oenology continues to evolve, the adoption of new irrigation technologies is reshaping the future of grape cultivation, paving the way for more sustainable and efficient practices in the vineyard.

Water conservation and its importance in viticulture

Water conservation plays a critical role in viticulture, as the availability of water directly impacts grape quality and yield. In recent years, the wine industry has increasingly recognized the importance of sustainable water management practices to ensure long-term viability. By implementing water-saving technologies such as drip irrigation systems and moisture sensors, vineyards can optimize water usage while minimizing waste. Additionally, the use of cover crops and mulching can help reduce evaporation and erosion, further conserving water in the vineyard. Beyond benefiting the environment, water conservation also contributes to economic sustainability by reducing water costs for vineyard operations. As climate change continues to pose challenges to water availability, embracing efficient water conservation practices will be essential for the future of viticulture and the overall success of the wine industry.

The effects of water stress on grape quality

Water stress is a significant factor that can influence grape quality, affecting various aspects of grape development and composition. Sustained water deficits can lead to decreased grape size, altered sugar accumulation, and changes in acidity levels, all of which can impact the overall quality of the grapes and the resulting wine. Research has shown that water stress can also affect the sensory characteristics of the wine, including aroma and flavor profiles. Furthermore, the timing and severity of water stress can play a crucial role in determining the extent of these effects. Understanding the effects of water stress on grape quality is essential for grape growers and winemakers to make informed decisions regarding irrigation strategies and vineyard management practices. By considering the impact of water stress on grape quality, practitioners can work towards producing high-quality wines that accurately reflect the terroir and climate of the vineyard.

VIII. SUSTAINABLE VITICULTURE

Sustainable viticulture represents a pivotal cornerstone in the ongoing revolution within the oenology and viticulture sphere. By prioritizing environmental conservation and biodiversity preservation, sustainable viticulture aims to minimize the impact of vineyard operations on the ecosystem while ensuring the long-term viability of grape cultivation. Through the integration of innovative practices such as organic farming, biodynamic viticulture, and precision agriculture, vintners can achieve a harmonious balance between quality wine production and ecological responsibility. Moreover, sustainable viticulture embodies a holistic approach that extends beyond the vineyard to encompass social and economic considerations, fostering a more sustainable and resilient industry. As the demand for environmentally-friendly practices continues to grow, sustainable viticulture is poised to play a significant role in shaping the future of oenology and viticulture, paving the way for a more sustainable and vibrant wine industry.

Principles of sustainability in wine production

One of the key principles of sustainability in wine production is the concept of environmental stewardship. This entails implementing practices that minimize environmental impact, such as reducing water usage, implementing renewable energy sources, and protecting biodiversity. By prioritizing sustainability in vineyard management and winemaking processes, producers can ensure the longevity of their operations and minimize their carbon footprint. Additionally, a focus on social equity is essential in sustainable wine production, ensuring fair labor practices and supporting local communities. Economic viability is also a crucial aspect, as sustainable practices can lead to cost savings in the long run. By incorporating these principles of sustainability into wine production, producers can not only create high-quality wines but also contribute to a more environmentally conscious and socially responsible industry. This shift towards sustainability is not only beneficial for the environment and society but also for the long-term success and reputation of wineries in the global market.

Organic and biodynamic farming practices

Organic and biodynamic farming practices have gained significant attention in the oenology and viticulture industry due to their emphasis on sustainability and environmental stewardship. These methods focus on holistic approaches to farming, utilizing natural inputs and avoiding synthetic chemicals. Organic farming promotes soil health and biodiversity, which can contribute to the overall quality of grapes and, consequently, the resulting wines. Biodynamic practices take this a step further by incorporating spiritual and astrological influences in farming activities. By treating the vineyard as a self-sustaining ecosystem, biodynamic farming aims to enhance the vitality of the soil and the health of the plants. While these practices require careful management and additional labor, their potential benefits in terms of grape quality, wine flavor, and long-term sustainability make them a compelling area for further research and implementation in the modern oenological landscape.

Certification programs and consumer awareness

Certification programs play a crucial role in enhancing consumer awareness in the field of oenology and viticulture. These programs provide a means for consumers to identify and differentiate products that meet specific quality standards and criteria. By obtaining certification, wineries and vineyards demonstrate their commitment to producing high-quality wines using sustainable and ethical practices. This not only helps to build trust with consumers but also educates them about the importance of supporting environmentally responsible and socially conscious producers. As consumers become more informed about certification programs and their significance, they are likely to make more informed purchasing decisions, ultimately driving demand for sustainably produced wines. By highlighting the value of certification programs in fostering consumer awareness, the oenology and viticulture industry can continue to evolve towards a more sustainable and ethical future.

IX. BIODIVERSITY IN VINEYARDS

Biodiversity in vineyards plays a crucial role in the sustainability and resilience of viticultural ecosystems. The diverse array of plant and animal species within vineyard environments contributes to pest control, soil health, and overall ecosystem balance. By promoting biodiversity through planting cover crops, preserving natural habitats, and avoiding monoculture practices, vineyard managers can enhance the long-term health of their vineyards and mitigate the need for chemical inputs. Research has shown that increased biodiversity in vineyards can lead to higher yields, improved grape quality, and enhanced ecosystem services. Therefore, integrating biodiversity conservation practices into vineyard management strategies is not only beneficial for the environment but also for the economic viability of vineyard operations. Moving forward, further research and adoption of biodiversity-enhancing techniques will be critical in ensuring the sustainability of vineyard ecosystems in the ever-evolving field of oenology and viticulture.

The role of biodiversity in ecosystem health

Biodiversity plays a crucial role in maintaining ecosystem health, particularly in the context of oenology and viticulture. The diverse array of species within an ecosystem provides essential services such as pollination, pest control, and soil nutrient cycling, all of which are integral to the health and productivity of vineyards. Research has shown that increasing biodiversity within vineyard landscapes can lead to improved pest management, enhanced soil quality, and greater overall resilience to environmental stresses. By promoting the presence of diverse plant and animal species within vineyard ecosystems, practitioners can cultivate more sustainable and regenerative agricultural practices that benefit both the environment and grape production. Embracing biodiversity in oenology and viticulture not only enhances the health of ecosystems but also contributes to the overall quality and longevity of wine production.

Strategies for promoting biodiversity

Strategies for promoting biodiversity within vineyards are essential in fostering a sustainable environment for oenology and viticulture. One approach involves integrating cover crops, such as legumes, grasses, or flowers, between vine rows to provide habitat for beneficial insects, improve soil health, and mitigate erosion. Additionally, implementing agroforestry practices by intercropping with trees or shrubs can enhance biodiversity by creating diverse habitats and increasing ecosystem resilience. Employing organic and biodynamic farming methods further supports biodiversity by reducing reliance on synthetic chemicals and preserving soil microbiota. These strategies not only contribute to a healthier environment but also enhance grape quality and vineyard sustainability in the long run. By promoting biodiversity in vineyards, oenologists and viticulturists can build more resilient ecosystems, improve wine quality, and contribute to the conservation of natural resources for future generations.

Benefits of biodiversity for disease control and yield

Biodiversity plays a crucial role in disease control and yield optimization in the context of oenology and viticulture. Increased biodiversity within vineyards can lead to a more balanced ecosystem, where natural predators of pests help reduce the need for chemical pesticides. This approach not only promotes environmental sustainability but also contributes to healthier vines and higher quality grapes. Additionally, diverse plant species can enhance soil health, nutrient cycling, and water retention, ultimately improving overall vineyard productivity. Research has shown that increased biodiversity can also lead to a more resilient ecosystem, better equipped to withstand climate variability and unexpected challenges. Embracing the benefits of biodiversity in vineyard management is essential for long-term sustainability and success in the evolving landscape of oenology and viticulture.

X. ADVANCES IN FERMENTATION TECHNOLOGY

Recent advances in fermentation technology have revolutionized the fields of oenology and viticulture, offering new possibilities for wine production and quality. Innovations in temperature-controlled fermentation tanks, the use of specific yeast strains, and the implementation of sensory analysis tools have significantly impacted the way wines are made. For example, the advent of micro-oxygenation has allowed winemakers to precisely control the amount of oxygen in the wine, leading to enhanced color stability and smoother tannins. Additionally, the emergence of enzymatic extraction techniques has improved the efficiency of maceration processes, resulting in wines with more intense flavors and aromas. These advancements in fermentation technology not only increase the quality of the final product but also allow winemakers to experiment with new styles and techniques, shaping the future of oenology and viticulture in exciting ways. As research in this area continues to evolve, the possibilities for innovation in wine production are endless.

Innovations in yeast and bacteria strains

In the realm of oenology and viticulture, advancements in yeast and bacteria strains have been pivotal in ushering in a new era of innovation. These strains play a crucial role in the fermentation process, ultimately influencing the flavor profile and quality of the final product. Researchers have been focusing on developing novel strains that can enhance aroma, mouthfeel, and overall complexity in wines. For example, genetically modified yeast strains have been engineered to produce specific aromatic compounds, leading to unique and desirable wine characteristics. Additionally, bacteria strains have been utilized for malolactic fermentation, contributing to the smoothness and stability of wines. These innovations have not only improved the efficiency of winemaking processes but have also expanded the range of flavor profiles that oenologists can achieve. As the field continues to evolve, the exploration of new yeast and bacteria strains holds great promise for the future of oenology and viticulture.

Controlled fermentation environments

Controlled fermentation environments play a crucial role in shaping the final quality and characteristics of wines. By closely monitoring and regulating variables such as temperature, oxygen levels, and nutrient availability, winemakers can exert greater control over the fermentation process. The use of temperature-controlled stainless steel tanks, for example, allows for precise temperature regulation, which can influence the rate of fermentation and the development of desirable aromas and flavors. Additionally, the introduction of inert gases like nitrogen can help to minimize oxidation and preserve the delicate nuances of the wine. These advancements in fermentation technology have not only improved the consistency and quality of wines but have also opened up new possibilities for experimentation and innovation in winemaking. As winemakers continue to refine and optimize these controlled fermentation environments, the future of oenology and viticulture holds promise for even greater advancements and discoveries.

The impact of fermentation technology on wine flavor profiles

Fermentation technology plays a pivotal role in shaping wine flavor profiles, contributing significantly to the complex aromas and tastes that define a wine's character. Through the controlled process of fermentation, yeasts interact with grape sugars, producing alcohol and a variety of compounds that influence the final product. The choice of yeast strain, fermentation temperature, and duration all impact the flavor development, with modern advancements allowing for greater precision and control. For example, the use of specific yeast strains that emphasize fruity or spicy notes can enhance the overall flavor profile of a wine. Furthermore, techniques such as cold soaking and extended maceration can extract more color and tannins, while malolactic fermentation can soften acidity and introduce buttery or creamy undertones. As oenologists continue to refine fermentation processes, the potential for creating unique and exceptional wine flavors becomes increasingly promising, opening new doors for innovation and creativity in the field of oenology.

XI. WINE AGING AND MATURATION

In the realm of wine production, the process of aging and maturation holds a significant place in crafting exceptional wines with complexity and character. XI. Wine Aging and Maturation plays a crucial role in defining the quality and flavor profile of a wine, as it allows the interplay of various chemical reactions to unfold over time. Factors such as oak aging, yeast autolysis, and bottle aging all contribute to the evolution of the wine, enhancing its aroma, taste, and texture. The art of wine aging requires precision and expertise to determine the optimal conditions and duration for maturation. Recent advancements in technology, such as controlled temperature and humidity environments, have enabled winemakers to better control the aging process and achieve desired results. Understanding and mastering the intricacies of wine aging and maturation is essential for producing exceptional wines that captivate the palate and stand the test of time in the ever-evolving world of oenology and viticulture.

New approaches to barrel aging

Recent advancements in oenology have seen a surge of interest in new approaches to barrel aging that are revolutionizing the industry. Traditional methods of barrel aging have long been a staple in winemaking, but now, innovative techniques are being explored to enhance the aging process and flavor profiles of wines. From the use of different types of wood, such as acacia or chestnut, to experimenting with varying toast levels and barrel sizes, winemakers are pushing boundaries to create unique and complex wines. Case studies have shown that these new approaches can result in wines with enhanced aromas, textures, and overall quality. As the field of oenology continues to evolve, these advancements in barrel aging are shaping the future of winemaking and offering exciting possibilities for the industry. The integration of new technologies and methodologies in barrel aging signifies a new era in oenology, promising a future full of innovative and exceptional wines.

Alternative aging materials and methods

The exploration of alternative aging materials and methods in oenology and viticulture marks a significant shift towards innovation and sustainability in the industry. Traditional practices have long relied on oak barrels for aging wine, but the limitations and environmental concerns associated with this approach have prompted researchers and winemakers to seek out alternatives. From concrete eggs to clay amphorae, a variety of materials offer unique characteristics that can impart distinct flavors and textures to wines. Additionally, innovative aging methods such as underwater aging or aging in tinajas have gained traction for their ability to elevate the complexity of wines while reducing the carbon footprint of production. By embracing these alternative materials and methods, the oenology and viticulture field is not only advancing its craft but also paving the way for a more sustainable and diverse future in wine production.

The science behind wine aging and its effects on quality

The science behind wine aging plays a vital role in determining the quality and characteristics of the final product. During the aging process, chemical reactions occur within the wine that influence its aroma, flavor, and texture. One key component in wine aging is the interaction between oxygen and phenolic compounds, which can lead to the development of complex flavors and aromas. Additionally, the presence of tannins in the wine can soften over time, resulting in a more harmonious and balanced taste. Understanding the mechanisms behind wine aging allows winemakers to manipulate these processes to achieve the desired flavor profile in their wines. Therefore, advancements in this field can lead to the production of high-quality wines with unique and appealing characteristics, ultimately shaping the future of oenology and viticulture in this revolutionary era.

XII. WINE CHEMISTRY AND ANALYTICS

In the realm of wine chemistry and analytics, advancements in technology and methodologies have greatly impacted the field of oenology and viticulture. The ability to accurately analyze the chemical composition of wine, such as sugar content, acidity, phenolic compounds, and volatile compounds, has revolutionized winemaking practices. From the use of high-performance liquid chromatography (HPLC) to mass spectrometry, these analytical tools provide winemakers with valuable insights into the quality and characteristics of their wines. Moreover, the integration of big data and artificial intelligence in wine analysis has further enhanced the precision and efficiency of wine production processes. By harnessing the power of cutting-edge analytics, oenologists and viticulturists can optimize their practices, improve wine quality, and cater to evolving consumer preferences. The fusion of chemistry and analytics is shaping a new era in oenology, promising innovative solutions and breakthroughs in the art of winemaking.

The role of chemistry in understanding wine

Chemistry plays a crucial role in unraveling the complexities of wine, making it an indispensable tool in the realm of oenology. By understanding the chemical composition of grapes and how it transforms during fermentation, researchers and winemakers can optimize processes to enhance the quality and characteristics of wine. From the influence of pH levels on flavor profiles to the role of organic compounds in aging potential, chemistry offers valuable insights into the science behind winemaking. Through the analysis of volatile compounds, tannins, and sugars, chemists can determine the unique fingerprint of each wine, elucidating its origin and quality. Utilizing spectroscopic methods and chromatography techniques, researchers can delve deeper into the molecular structure of wine, providing a more nuanced understanding of its complexity. Ultimately, chemistry serves as a cornerstone in comprehending the intricacies of wine, paving the way for innovation and advancement in oenology.

Advanced analytical techniques for wine composition

Advanced analytical techniques for wine composition have revolutionized the way oenologists and viticulturists understand and manipulate the complex factors that contribute to the final product. High-resolution mass spectrometry, nuclear magnetic resonance spectroscopy, and chromatographic methods have provided unparalleled insights into the chemical makeup of wines, enabling precise identification of compounds responsible for flavor, aroma, color, and aging potential. These cutting-edge tools allow researchers to not only analyze individual components but also to study their interactions and synergies, leading to a more holistic comprehension of wine composition. By leveraging these advanced techniques, winemakers can optimize their production processes, enhance quality control measures, and even innovate new varieties that cater to evolving consumer preferences. As this new era of oenology and viticulture unfolds, the integration of advanced analytical techniques promises to shape the future of the industry by pushing boundaries, inspiring creativity, and setting new standards of excellence.

Applications of chemometrics in oenology

Chemometrics, a powerful tool that combines statistics, mathematics, and chemistry, has found significant applications in oenology. By analyzing complex data sets from various stages of winemaking, chemometrics enables researchers and winemakers to optimize processes, control quality, and make informed decisions in vineyard management. From grape selection to fermentation and aging, chemometrics plays a crucial role in understanding the chemical composition of wines, identifying key compounds, and predicting sensory attributes. Techniques like cluster analysis, principal component analysis, and partial least squares regression have been instrumental in characterizing wines, detecting adulteration, and even determining regional origins. The integration of chemometrics in oenology not only enhances the efficiency and precision of winemaking but also opens up new possibilities for innovation and experimentation in the field. As the wine industry embraces technological advancements, the application of chemometrics continues to revolutionize the production and appreciation of wines in this new era of oenology and viticulture.

XIII. SENSORY SCIENCE AND CONSUMER PREFERENCES

In the realm of oenology, the marriage of sensory science and consumer preferences has become a paramount consideration in the production and marketing of wines. With the advent of advanced sensory evaluation techniques and research methodologies, winemakers are now able to delve deeper into understanding the sensory profiles of their wines and how they resonate with consumers. By incorporating consumer preferences into the winemaking process, producers can tailor their products to meet market demands more effectively. This integration of sensory science and consumer insights has not only enhanced the quality and appeal of wines but has also enabled winemakers to create unique and innovative products that stand out in a competitive market. As this synergy continues to evolve, the future of oenology holds promise for a more consumer-focused and sensorially driven approach to winemaking.

The study of sensory perception in wine tasting

The study of sensory perception in wine tasting plays a crucial role in understanding the complex interactions between human senses and wine characteristics. By delving into the intricate nuances of flavors, aromas, textures, and colors, researchers can unravel the mysteries behind why certain wines evoke specific sensory responses in consumers. Utilizing advanced sensory analysis techniques, such as descriptive sensory analysis and consumer testing, oenologists can glean valuable insights into the preferences and perceptions of wine enthusiasts. Furthermore, incorporating cutting-edge technologies like electronic noses and tongues enables researchers to detect subtle differences in wine composition that are imperceptible to the human palate. By leveraging these tools and methodologies, oenologists can enhance the quality and consistency of wine production, ultimately elevating the overall sensory experience for consumers. The study of sensory perception in wine tasting heralds a new era of precision and innovation in oenology, shaping the future of wine production and appreciation.

The influence of sensory attributes on consumer choices

Consumer choices in the field of oenology are heavily influenced by sensory attributes, playing a crucial role in the selection of wines. The sensory experience encompasses a range of factors, including aroma, taste, color, and texture, all of which contribute to the overall perception of a wine. Studies have shown that consumers are not only drawn to wines with appealing sensory profiles but are also willing to pay a premium for wines that are perceived as high-quality in terms of their sensory attributes. Winemakers and viticulturists must therefore pay close attention to the sensory characteristics of their products, as these can significantly impact consumer preferences and purchasing decisions. By understanding and harnessing the power of sensory attributes, producers can create wines that stand out in a competitive market, ultimately driving sales and establishing a strong brand presence.

The development of sensory profiles for wine branding

By understanding the intricate sensory characteristics of different wines, producers can effectively communicate the unique aspects of their products to consumers. Recent advancements in sensory analysis techniques, such as aroma profiling and flavor profiling, have allowed for a more nuanced understanding of how various factors influence the sensory experience of wine. Utilizing these sensory profiles, wine brands can differentiate themselves in a crowded market, build strong brand identities, and create deeper connections with consumers. Additionally, sensory profiles can provide valuable insights for improving winemaking techniques, optimizing production processes, and ensuring consistency in quality. As such, the development of sensory profiles for wine branding represents a significant advancement in the field, shaping the future of oenology and viticulture.

XIV. WINE PACKAGING INNOVATIONS

The evolution of wine packaging has been a crucial area of innovation within the oenology field, with the XIV. Wine Packaging Innovations marking a significant advancement in this domain. This chapter explores the latest trends and developments in wine packaging, such as the use of sustainable materials, lightweight bottles, and innovative closures. These innovations not only enhance the aesthetic appeal of wine bottles but also play a vital role in preserving the quality and longevity of the wine. Case studies and research data will be presented to illustrate the impact of these packaging innovations on consumer perception and market competitiveness. By embracing these advancements, wineries can differentiate themselves in the market and cater to the evolving preferences of consumers. The chapter will conclude by reflecting on the future possibilities and challenges in wine packaging, outlining potential directions for further research and development in this dynamic field.

Eco-friendly and sustainable packaging solutions

The wine industry is undergoing a significant transformation towards more sustainable practices, including the adoption of eco-friendly packaging solutions. With the growing awareness of environmental issues, wineries are exploring new ways to reduce their carbon footprint and minimize waste. One approach gaining traction is the use of biodegradable and compostable materials for packaging, such as plant-based plastics or recycled paper. These alternatives not only help in reducing the reliance on non-biodegradable materials but also contribute to lowering energy consumption during production and transportation. By implementing these sustainable packaging solutions, wineries can align their practices with consumer preferences for environmentally conscious products while also demonstrating a commitment to long-term environmental stewardship. Moving forward, further research and innovation in this area can lead to more efficient and effective eco-friendly packaging options, solidifying the wine industry's role in promoting sustainability.

The evolution of wine bottle design and materials

The evolution of wine bottle design and materials has played a significant role in the advancement of oenology and viticulture. From traditional glass bottles to innovative materials such as lightweight plastics and sustainable alternatives like recycled glass, the industry has seen a profound shift in how wine is packaged and consumed. The aesthetic appeal of wine bottles has also evolved, with unique shapes, colors, and labeling techniques becoming more prevalent. Furthermore, advances in packaging technology have led to improvements in preserving the wine's quality and aging potential. As consumers become more environmentally conscious, there is a growing demand for eco-friendly packaging solutions. The future of wine bottle design and materials is likely to continue on a path of sustainability and innovation, reflecting the industry's commitment to both quality and environmental responsibility.

The impact of packaging on wine preservation and sales

Packaging plays a crucial role in the preservation and sales of wine, impacting both the quality of the product and consumer perception. Innovative packaging solutions such as advanced glass bottles, can coatings, and sustainable materials have the potential to prolong the shelf life of wines, preserving their flavor and aroma. Studies have shown that the design and material of the packaging can influence consumer purchasing decisions, with sleek and eco-friendly options often preferred. By incorporating cutting-edge packaging techniques, wineries can differentiate themselves in a competitive market, attracting environmentally conscious consumers and enhancing brand image. The strategic use of packaging can also contribute to increased sales and brand loyalty, as it communicates value and quality to consumers. As the wine industry continues to evolve, the significance of packaging in wine preservation and sales cannot be underestimated.

XV. WINE MARKETING AND DIGITAL TRANSFORMATION

In the realm of wine marketing, the landscape is undergoing a profound transformation catalyzed by the rapid integration of digital technologies. The evolution of consumer behavior and preferences has necessitated a shift towards innovative strategies that leverage the power of digital platforms to engage with a diverse and discerning audience. From targeted social media campaigns to personalized email marketing, the digital realm offers unparalleled opportunities for wine producers to connect with consumers in a more direct and impactful manner. By harnessing the vast array of data analytics tools and artificial intelligence capabilities, wineries can better understand consumer sentiments, preferences, and trends, enabling them to tailor their marketing efforts with precision and efficiency. This digital revolution in wine marketing is not only revolutionizing how products are promoted and sold but also reshaping the very fabric of the industry, paving the way for a more dynamic and interactive relationship between producers and consumers.

The role of digital marketing in the wine industry

Digital marketing plays a pivotal role in the wine industry, offering wineries a platform to reach a wider audience and engage with consumers in innovative ways. Through targeted online campaigns, social media strategies, and interactive websites, wineries can enhance brand visibility, build customer relationships, and drive sales. By analyzing consumer data and behavior patterns, digital marketing allows wineries to tailor their messaging and product offerings to meet the evolving demands of the market. Case studies have shown the positive impact of digital marketing on the success of wine brands, with increased sales, brand loyalty, and market share. As the wine industry continues to adapt to changing consumer preferences and market dynamics, the integration of digital marketing strategies will be crucial for wineries to remain competitive and relevant in this new era of oenology and viticulture.

E-commerce and direct-to-consumer sales strategies

E-commerce and direct-to-consumer sales strategies have emerged as pivotal components in the revolution of oenology and viticulture. With the rise of online platforms and digital marketing, wineries are now able to reach a broader audience and establish direct connections with consumers. By bypassing traditional distribution channels, wineries can increase their profits and cultivate brand loyalty through personalized interactions. Moreover, these strategies enable wineries to gather valuable data on consumer preferences and behavior, allowing for tailored marketing campaigns and product development. Case studies have shown that wineries implementing e-commerce and direct-to-consumer sales have experienced significant growth and enhanced customer engagement. As the industry continues to evolve, leveraging these strategies will be essential for wineries to stay competitive and thrive in the new era of oenology and viticulture.

Leveraging social media and technology for brand building

In today's competitive marketplace, leveraging social media and technology has become essential for brand building in the oenology and viticulture industries. By utilizing platforms such as Instagram, Facebook, and Twitter, wineries and vineyards can reach a wider audience, engage with consumers on a personal level, and showcase their unique brand identity. Social media allows for real-time interactions, instant feedback, and the opportunity to build a loyal community of wine enthusiasts. Furthermore, technology such as augmented reality apps, virtual tours, and online wine tastings can enhance the consumer experience, educate customers about the winemaking process, and ultimately drive sales. By embracing these digital tools, oenologists and viticulturists can stand out in a crowded market, build strong brand recognition, and remain at the forefront of this revolutionary era in the wine industry.

XVI. GLOBAL WINE TRADE AND ECONOMICS

Global Wine Trade and Economics chapter delves into the intricacies of the wine market on a worldwide scale, highlighting the economic significance of the industry. With the rise of globalization and the increasing popularity of wine consumption, understanding the dynamics of global trade is crucial for both producers and consumers. This chapter investigates the role of trade agreements, tariffs, and market trends in shaping the wine economy, shedding light on the challenges and opportunities faced by different regions. By analyzing data on exports, imports, and pricing, this chapter aims to provide a comprehensive overview of the current state of the global wine trade and its implications for the industry. As wine continues to transcend borders and cultures, the economic aspects of production, distribution, and consumption play a pivotal role in shaping the future of oenology and viticulture.

The current state of the global wine market

The current state of the global wine market showcases a complex landscape shaped by evolving consumer preferences, changing climate patterns, and advancements in technology. The demand for premium and organic wines is on the rise, prompting winemakers to invest in sustainable practices and innovative solutions to meet these demands. In addition, the emergence of new wine-producing regions, such as China and India, is altering traditional market dynamics and creating opportunities for growth and diversification. The use of precision viticulture techniques, artificial intelligence, and data analytics is revolutionizing grape cultivation and winemaking processes, enhancing quality and efficiency. As we navigate this new era of oenology and viticulture, it is crucial to adapt to these changes, while also anticipating future trends and challenges to ensure the sustainability and competitiveness of the industry.

Trade agreements and their impact on viticulture

Trade agreements play a crucial role in the global landscape of viticulture, directly impacting the production, distribution, and consumption of wines. These agreements, which regulate the terms of trade between countries, can either facilitate or hinder the growth of the wine industry. By reducing tariffs and trade barriers, trade agreements enable winemakers to access new markets, increase export opportunities, and expand their businesses. On the other hand, restrictions or changes in trade agreements can disrupt established trade relationships, affecting the competitiveness of wine producers. Understanding the implications of trade agreements on viticulture is essential for navigating the complexities of the international wine market. As the wine industry continues to evolve in response to various economic and political factors, analyzing the effects of trade agreements becomes increasingly important for shaping the future of oenology and viticulture.

Economic challenges and opportunities in the wine industry

One of the key aspects of the wine industry that has significant economic implications is the balance between challenges and opportunities. The economic challenges in the wine industry often revolve around fluctuating market demands, production costs, and global competition. These challenges can put pressure on wineries to optimize their operations, adapt to changing consumer preferences, and implement cost-effective strategies to remain competitive. However, within these challenges lie opportunities for growth and innovation. By leveraging advancements in technology, sustainability practices, and marketing strategies, wineries can capitalize on emerging trends and tap into new markets. For example, the rise of e-commerce has provided wineries with a direct channel to reach consumers and expand their customer base. Ultimately, navigating these economic challenges and seizing opportunities in the wine industry requires a strategic approach that balances risk and reward to drive long-term success.

XVII. WINE EDUCATION AND SOMMELIER SCIENCE

Wine Education and Sommelier Science, the importance of knowledge and expertise in wine tasting and selection cannot be overstated. Today, the role of a sommelier extends beyond merely recommending wines to diners; it involves a deep understanding of grape varieties, regions, production methods, and food pairings. As such, wine education has become an indispensable part of sommelier training, with courses covering sensory evaluation, wine production, and service techniques. These programs not only cultivate the refined palates of aspiring sommeliers but also equip them with the skills to navigate the complex and ever-evolving world of wine. Moreover, the fusion of science and art in sommelier training is ushering in a new era of wine appreciation, where individuals are not only consumers but connoisseurs in their own right. As the field of wine education continues to evolve, the role of sommeliers will undoubtedly become more pivotal in shaping the future of oenology and viticulture.

The evolution of wine education programs

The evolution of wine education programs has played a crucial role in shaping the modern landscape of oenology and viticulture. As the demand for higher quality wines and skilled professionals continues to rise, educational institutions around the world have been adapting their programs to meet these needs. In recent years, there has been a shift towards more comprehensive and hands-on training, with a focus on practical skills, sensory analysis, and sustainable practices. This shift reflects the industry's recognition of the importance of a well-rounded education that integrates traditional knowledge with cutting-edge research and technology. By offering students opportunities to learn from experienced professionals, visit vineyards and wineries, and participate in research projects, these programs are equipping future winemakers and viticulturists with the tools they need to succeed in this dynamic and competitive field. The evolution of wine education programs is not only preparing students for the challenges of today but also laying the foundation for a more innovative and sustainable future in oenology and viticulture.

The science behind sommelier training

The science behind sommelier training is a complex and multi-faceted process that encompasses the study of wine production, grape varieties, regions, tasting techniques, food pairing, and much more. Sommeliers undergo rigorous training to develop their sensory skills, wine knowledge, and ability to effectively communicate with patrons. This training often involves blind tastings, where sommeliers must identify wines based on taste, aroma, and appearance. Additionally, sommeliers must have a deep understanding of viticulture and oenology to be able to navigate the vast and constantly evolving world of wine. By studying the science behind winemaking, sommeliers can better appreciate the nuances of different styles and regions, allowing them to provide expert recommendations and enhance the dining experience for their guests. The combination of scientific knowledge and sensory expertise is what sets sommeliers apart as true wine professionals in the industry.

The future of wine expertise and certification

The future of wine expertise and certification is undergoing a transformation in response to the evolving landscape of oenology and viticulture. With advancements in technology, the traditional ways of certifying wine experts are being challenged, leading to a reevaluation of the certification process. As the industry becomes more specialized and nuanced, there is a growing need for a more dynamic and tailored approach to certifying wine professionals. This includes integrating new tools such as artificial intelligence and data analytics to assess expertise and proficiency accurately. Furthermore, there is a shift towards a more global perspective, recognizing the diversity of wine regions and styles worldwide. The future of wine expertise and certification will likely involve a more interdisciplinary and multifaceted approach, taking into account not only technical knowledge but also sensory skills, cultural understanding, and sustainability practices. This new era of certification will ensure that wine professionals are equipped to navigate the complexities of the industry and contribute to its continued innovation and development.

XVIII. WINE TOURISM AND EXPERIENCE ECONOMY

The intersection of wine tourism and the experience economy in the XVIII century has sparked a new wave of innovation in the fields of oenology and viticulture. As consumers increasingly seek unique and immersive experiences, wineries are adapting by offering not just tastings, but full-fledged experiences that engage all the senses. From vineyard tours and hands-on harvesting activities to food and wine pairings, these experiences are reshaping the way people engage with wine. Case studies have shown that wineries incorporating experiential elements into their offerings are not only attracting more visitors but also fostering greater brand loyalty and long-term relationships with customers. This shift towards experiential wine tourism signifies a fundamental change in the industry, where the focus is no longer solely on the product but on the holistic experience created around it. In this new era, wineries must continue to innovate and adapt to meet the demands of discerning consumers seeking memorable and transformative experiences, setting the stage for a future where wine tourism is an integral part of the overall wine industry landscape.

The growth of wine tourism

Wine tourism has witnessed significant growth in recent years, driven by an increasing demand for immersive experiences in the world of oenology and viticulture. This trend is reshaping the traditional wine industry by fostering closer relationships between producers and consumers, and by offering visitors a firsthand look at the winemaking process. The rise of wine tourism has not only boosted sales and brand recognition for wineries but has also fueled local economies through the development of hospitality services, restaurants, and related businesses. Furthermore, it has encouraged environmental sustainability practices in the industry as consumers become more aware of the impact of their choices. As the wine tourism sector continues to expand, it presents new opportunities for research and innovation in enhancing visitor experiences, preserving cultural heritage, and promoting sustainable practices in the production and consumption of wine.

Creating immersive experiences for wine enthusiasts

Creating immersive experiences for wine enthusiasts has become a pivotal strategy in engaging consumers and fostering a deep appreciation for the art of winemaking. By leveraging virtual reality (VR) and augmented reality (AR) technologies, wineries can transport visitors to vineyards, cellars, and tasting rooms from the comfort of their own homes. These immersive experiences allow wine enthusiasts to virtually tour estates, witness the winemaking process up close, and even participate in interactive tasting sessions guided by experts. This innovative approach not only enhances customer engagement but also provides valuable insights into consumer preferences and behaviors. By offering these meticulously crafted experiences, wineries can connect with their audience on a more personal level, ultimately cultivating a loyal following and driving sales. As the wine industry continues to embrace technological advancements, the potential for creating truly unforgettable and educational experiences for wine enthusiasts is limitless.

The economic impact of wine tourism on regions

Wine tourism has emerged as a significant contributor to the economic development of regions, offering a unique opportunity for growth and diversification. The influx of visitors to wineries and vineyards not only boosts local economies but also creates jobs, stimulates related industries, and enhances the overall tourism sector. By showcasing the rich cultural heritage and natural beauty of wine-producing regions, wine tourism attracts a diverse range of tourists seeking unique experiences and premium products. Through guided tours, tastings, and events, wineries can directly engage with consumers and increase sales, thus generating additional revenue. Moreover, the promotion of sustainable practices in wine tourism can lead to environmental benefits and long-term economic sustainability for regions. As a result, the economic impact of wine tourism on regions is profound, highlighting the importance of leveraging this growing trend in the global wine industry.

XIX. REGULATORY CHANGES AND WINE LAW

Regulatory Changes and Wine Law, it is evident that the landscape of oenology and viticulture is undergoing a significant transformation. As new technologies and techniques are being implemented, there is a pressing need to adapt existing regulatory frameworks to ensure the industry's continued growth and sustainability. The evolution of wine law is crucial in this new era of revolutionary oenology, as it sets the guidelines for production, labeling, and marketing practices. By examining the impact of regulatory changes on the industry, we can gain insights into how these developments shape the future of oenology and viticulture. This analysis also raises important questions about the potential challenges and opportunities that lie ahead, prompting further research and exploration in this dynamic field.

Recent changes in wine legislation

Recent changes in wine legislation have significantly influenced the landscape of oenology and viticulture. Governments around the world are implementing new regulations to address issues such as sustainability, labeling requirements, and production standards. For instance, the European Union has introduced stricter guidelines on additives and pesticides in winemaking, aiming to promote environmental sustainability and consumer safety. These legislative changes have forced winemakers to adapt their practices, leading to a shift towards organic and biodynamic viticulture. Moreover, some regions have established specific designations of origin to protect traditional winemaking practices and maintain the unique characteristics of their wines. As a result, these recent legislative developments are reshaping the way wine is produced, marketed, and consumed, highlighting the importance of staying informed and compliant in the ever-evolving wine industry.

The impact of regulations on production and distribution

Regulations play a crucial role in shaping the production and distribution of wines in the oenology and viticulture industry. These regulations are put in place to ensure the quality and safety of wines for consumers, as well as to protect the environment and preserve the heritage of winemaking regions. From labeling requirements to restrictions on additives and pesticides, regulations can significantly impact the way wines are produced and marketed. For producers, navigating these regulations can be a complex process that requires careful attention to detail and compliance with various standards. However, regulations also serve to standardize practices and ensure that wines meet certain quality standards, ultimately benefiting both producers and consumers alike. By understanding and adapting to these regulations, producers can innovate and improve their techniques while maintaining the integrity of their products in this new era of oenology and viticulture.

International wine laws and their harmonization

In the realm of oenology and viticulture, the importance of international wine laws cannot be understated. These regulations serve as a crucial framework for ensuring quality standards, protecting consumers, and maintaining the integrity of the wine industry. However, with the growing globalization of the wine market, there is an increasing need for the harmonization of these laws across different countries. This harmonization is essential to facilitate trade, reduce barriers, and promote a level playing field for wine producers around the world. By aligning regulations on issues such as labeling, production methods, and geographic indications, international wine laws can contribute to a more efficient and transparent global wine market. Therefore, efforts to harmonize these laws are vital in promoting innovation and sustainability within the industry, ultimately shaping the future of oenology and viticulture on a global scale.

XX. WINE AND HEALTH

Wine and health have been subjects of longstanding interest, with research increasingly shedding light on the potential benefits of moderate wine consumption. Studies have indicated that certain components in wine, such as polyphenols and resveratrol, may offer protective effects against cardiovascular diseases, inflammation, and oxidative stress. These findings have led to a growing body of literature supporting the notion that moderate wine consumption, particularly red wine, can be part of a healthy lifestyle. However, it is crucial to note that excessive alcohol intake can have detrimental effects on health, highlighting the importance of responsible drinking practices. As oenology and viticulture continue to advance, ongoing research into the potential health benefits of wine consumption will play a significant role in shaping consumer perceptions and industry practices, thus underscoring the need for further exploration and understanding in this area.

Research on wine's health benefits and risks

Wine has long been the subject of research regarding its potential health benefits and risks. Studies have shown that moderate wine consumption, particularly red wine, may have positive effects on cardiovascular health due to the presence of antioxidants like resveratrol. These compounds have been linked to reduced inflammation and improved blood vessel function. However, it is essential to note that excessive alcohol consumption can have detrimental effects on one's health, including an increased risk of liver disease, high blood pressure, and certain types of cancer. Researchers are continuously exploring the delicate balance between the potential benefits and risks of wine consumption, striving to provide evidence-based recommendations for the public. By understanding the complexities of wine's impact on health, future studies can help individuals make informed decisions about their consumption habits, guiding public health policies and initiatives.

The role of moderation in wine consumption

Moderation in wine consumption plays a pivotal role in promoting health benefits while also avoiding potential risks associated with excessive alcohol intake. Research has shown that moderate wine consumption, particularly red wine, can have positive effects on cardiovascular health, thanks to its high concentration of antioxidant compounds like resveratrol and polyphenols. However, it is crucial to emphasize the importance of moderation in wine consumption to reap these benefits without crossing the threshold into harmful territory. Guidelines for moderate consumption vary by individual factors such as age, gender, weight, and overall health status, underscoring the need for personalized recommendations in this context. By understanding and promoting moderation in wine consumption, oenologists and viticulturists can contribute to a holistic approach to wine enjoyment that prioritizes health and well-being while also embracing the cultural and sensory aspects of wine appreciation.

Wine in the context of dietary guidelines

In the context of dietary guidelines, wine has long been a subject of debate. While some studies have suggested that moderate wine consumption may have health benefits, particularly in terms of cardiovascular health due to its potential antioxidant properties and resveratrol content, others caution against excessive alcohol intake for various health reasons. Dietary guidelines generally recommend moderation in alcohol consumption, including wine, to reduce the risk of negative health outcomes such as liver disease, addiction, and certain cancers. The key lies in understanding the balance between the potential benefits and risks associated with wine consumption, and how it fits into an overall healthy diet. As oenology and viticulture continue to evolve with new advancements and innovations, it is crucial to consider the implications of these changes on dietary recommendations and public health messaging surrounding wine consumption. This intersection highlights the importance of ongoing research and collaboration between scientific fields to inform evidence-based guidelines for the public.

XXI. THE ROLE OF WOMEN IN OENOLOGY

In the XXI century, the role of women in oenology has gained prominence, marking a significant shift in the traditionally male-dominated field. Women have increasingly taken on key positions in wineries, vineyards, and research institutions, contributing unique perspectives and expertise to the industry. Their influence can be seen in the development of innovative techniques, sustainable practices, and a growing emphasis on terroir-driven winemaking. Studies have shown that women bring a different approach to wine production, focusing on precision, creativity, and attention to detail. Case studies of successful female winemakers have demonstrated their ability to produce high-quality wines that reflect both tradition and innovation. As more women enter the field of oenology, their contributions will continue to shape the future of winemaking, paving the way for new discoveries and advancements in viticulture.

Historical contributions of women in winemaking

Throughout history, women have made significant contributions to the art and science of winemaking, often overshadowed by their male counterparts. In ancient civilizations such as Egypt, Greece, and Rome, women played essential roles in cultivating grapes, fermenting wines, and even serving as priestesses dedicated to wine deities. Fast forward to the Middle Ages, women continued to influence winemaking practices, particularly in monastic settings where they were tasked with managing vineyards and producing wines for religious ceremonies. Despite these historical contributions, the role of women in winemaking has been undervalued and underrepresented. However, with the rise of gender equality and recognition for women in the wine industry, there is a newfound appreciation for their expertise and creativity in shaping the future of oenology and viticulture. The revolutionary era of winemaking acknowledges and celebrates the historical contributions of women, paving the way for greater diversity and innovation in the field.

Current trends in female leadership in the wine industry

Current trends in female leadership in the wine industry reflect a progressive shift towards gender equality and empowerment within a traditionally male-dominated field. Women are increasingly occupying key positions as winemakers, vineyard managers, sommeliers, and executives, bringing unique perspectives and skills to the industry. As more women enter the wine sector, they are challenging norms and stereotypes, fostering innovation, and driving change. Studies have shown that diverse leadership teams, including more female representation, are linked to better financial performance and decision-making. Female leaders in the wine industry are paving the way for future generations of women to thrive in oenology and viticulture. By promoting inclusivity and advocating for equality, these women are shaping a more diverse and dynamic industry that is better equipped to meet the challenges and opportunities of the future.

The future of gender diversity in oenology and viticulture

Gender diversity in oenology and viticulture is a topic that has gained increasing attention in recent years. Historically, the wine industry has been predominantly male-dominated, with few opportunities for women to enter or advance in the field. However, as the industry undergoes a period of transformation and innovation, there is a growing recognition of the importance of diversity and inclusivity. Initiatives aimed at increasing gender diversity in oenology and viticulture, such as mentorship programs, scholarships, and networking events, are beginning to gain traction. With more women entering the field and taking on leadership roles, the future of oenology and viticulture looks increasingly diverse and dynamic. Moving forward, it will be essential for the industry to continue supporting and promoting gender diversity, as a more inclusive workforce will bring new perspectives, ideas, and innovations to the field.

XXII. EMERGING WINE REGIONS

The exploration of emerging wine regions represents a significant aspect of the ongoing revolutionary era in oenology and viticulture. These regions, often characterized by their unique terroir and climate conditions, offer exciting opportunities for experimentation and innovation in winemaking. By diversifying the global wine map, these emerging regions challenge the traditional notions of wine production and consumption, pushing boundaries and expanding the horizons of the industry. One such example is the rise of Eastern European countries like Hungary and Bulgaria, which are gaining recognition for their distinctive varietals and high-quality wines. The exploration and development of these new frontiers not only showcase the adaptability and resilience of the industry but also hint at the potential for exciting discoveries and new trends to shape the future of oenology and viticulture.

Characteristics of new and promising wine regions

Characteristics of new and promising wine regions are vital in understanding the changing landscape of oenology and viticulture. These regions exhibit unique terroirs, which contribute to the distinct flavor profiles of the wines produced. The climate, soil composition, elevation, and proximity to bodies of water all play a crucial role in determining the quality and style of wines from these regions. Additionally, the adoption of sustainable practices and technological innovations are prevalent in these emerging wine regions, allowing for the production of high-quality wines while minimizing environmental impact. The growing interest in these new regions is evident in the increasing number of awards and accolades received by their wines, signaling a shift in consumer preferences towards more diverse and adventurous wine choices. As these regions continue to gain recognition and popularity, they are poised to become driving forces in the future of oenology and viticulture, pushing the boundaries of traditional winemaking techniques and expanding the diversity of the global wine market.

Challenges faced by emerging wine producers

Emerging wine producers face a myriad of challenges as they navigate the competitive and ever-evolving landscape of the wine industry. One major hurdle is the lack of resources and capital needed to invest in innovative techniques and technologies that can enhance the quality and efficiency of wine production. Additionally, emerging producers often struggle to establish a strong brand presence and compete with well-established wineries that have a loyal customer base. Furthermore, the changing climate conditions and environmental challenges pose a threat to the sustainability of vineyards, requiring producers to adapt and implement resilient practices. Despite these obstacles, emerging wine producers have the opportunity to leverage their agility and creativity to differentiate themselves in the market and carve out a niche for their unique offerings. By staying abreast of industry trends and embracing innovation, emerging producers can overcome these challenges and thrive in the dynamic world of wine production.

The potential impact of new regions on the global wine scene

The potential impact of new regions on the global wine scene is a topic of great interest and significance in the field of oenology and viticulture. As traditional wine-producing regions face challenges such as climate change and increasing competition, emerging regions are gaining attention for their unique terroir and quality wines. These new regions offer opportunities for experimentation with different grape varieties, techniques, and styles, leading to a diversification of the global wine market. For example, regions like Oregon in the United States, Tasmania in Australia, and Ningxia in China have been attracting recognition for their distinct wines and are poised to make significant contributions to the industry. By exploring these new regions and their potential, researchers and winemakers can expand their knowledge and techniques, ultimately shaping the future landscape of oenology and viticulture.

XXIII. URBAN WINERIES AND MICRO-VITICULTURE

Urban wineries and micro-viticulture represent a significant shift in the traditional landscape of oenology and viticulture. These innovative approaches bring winemaking closer to urban areas, thereby reducing transportation costs, increasing accessibility for consumers, and promoting sustainable practices. By utilizing small plots of land within cities, micro-viticulture allows for more precise and experimental grape growing techniques that can lead to unique and high-quality wines. Urban wineries, on the other hand, provide a space for winemaking facilities in bustling metropolitan areas, attracting a new demographic of wine enthusiasts and establishing a connection between urban dwellers and the winemaking process. This trend towards urban wineries and micro-viticulture reflects a growing demand for artisanal, locally-produced wines, and signals a promising future for the industry as it continues to evolve in response to changing consumer preferences and environmental concerns.

The rise of urban winemaking

The rise of urban winemaking has emerged as a notable trend within the field of oenology, marking a significant departure from traditional practices. This phenomenon signifies a shift towards innovation and experimentation, as winemakers seek to create unique and distinctive wines in non-traditional settings. Urban winemaking has become a focal point for research and development, as it offers a platform for exploring new techniques and methodologies that challenge the status quo. By engaging with urban environments, winemakers are able to adapt to changing climates and consumer preferences, while also fostering a sense of community and sustainability. As this trend continues to gain momentum, it is clear that urban winemaking is poised to shape the future of oenology and viticulture, opening up new possibilities for exploration and growth in the field.

Techniques and challenges of micro-viticulture

In the realm of viticulture, micro-viticulture has emerged as a revolutionary technique that allows for precision and customization in grape growing. This approach involves meticulously managing every aspect of the vineyard on a small scale, from soil composition to irrigation practices. By tailoring these factors to meet the specific needs of each vine, micro-viticulture can optimize grape quality and yield. However, this level of detail also presents challenges, such as increased labor and resource requirements. To successfully implement micro-viticulture, vineyard managers must have a deep understanding of the intricacies of grape growing and be willing to invest time and effort into monitoring and adjusting various parameters. Despite these challenges, the potential benefits of micro-viticulture are undeniable, with the promise of producing higher quality grapes and wines that reflect the unique terroir of a vineyard.

The role of urban wineries in local economies

Urban wineries play a crucial role in local economies by contributing to job creation, tourism, and economic growth. These establishments offer unique opportunities for urbanites to experience the winemaking process firsthand, establishing a connection between consumers and producers. By sourcing grapes from nearby vineyards and producing wines within city limits, urban wineries support local agriculture and promote sustainability. Additionally, these wineries often serve as cultural hubs, hosting events, tastings, and educational programs that attract both locals and tourists, further stimulating the local economy. The presence of urban wineries can revitalize neighborhoods, attract investment, and showcase the potential for innovation in the wine industry. As such, the growth of urban wineries is not only a reflection of changing consumer preferences but also a significant contributor to the economic vitality of urban areas.

XXIV. WINE AND TECHNOLOGY: THE ROLE OF AI

In the current landscape of oenology and viticulture, the integration of artificial intelligence (AI) has emerged as a game-changer in the wine industry. AI has revolutionized the way winemakers approach various aspects of the winemaking process, from vineyard management to fermentation control and quality assessment. By harnessing AI technology, winemakers can optimize yield, monitor grape maturity, predict wine quality, and even personalize recommendations for consumers. Case studies have shown that AI-driven systems can accurately predict optimal harvest times, resulting in higher-quality wines and more efficient production processes. The potential of AI in the wine industry is vast, with advancements in machine learning algorithms continuously pushing the boundaries of what is possible. As we delve deeper into this new era of oenology and viticulture, the role of AI will undoubtedly play a pivotal role in shaping the future of winemaking, offering exciting opportunities for innovation and growth.

Artificial intelligence in wine pattern recognition

Recent advancements in artificial intelligence have opened up exciting new possibilities in the field of wine pattern recognition. By utilizing machine learning algorithms and neural networks, researchers and oenologists can now analyze vast amounts of data to identify intricate patterns and trends in the characteristics of various wines. This technology has the potential to revolutionize the way wines are classified, labeled, and even produced. With the ability to detect subtle flavor profiles, tannin levels, and aging potential, artificial intelligence can assist in quality control, blending decisions, and even predicting market trends. By incorporating these cutting-edge technologies into the traditional practices of oenology and viticulture, we are entering a new era of precision and innovation that has the power to transform the industry. As we continue to explore the applications of artificial intelligence in wine analysis, the future of oenology and viticulture looks promising, with endless possibilities for growth and advancement.

AI-driven decision-making in viticulture

AI-driven decision-making in viticulture has emerged as a revolutionary tool to enhance grape growing and winemaking processes. By utilizing advanced technologies such as machine learning and data analytics, AI systems can analyze vast amounts of data from vineyards, weather patterns, soil conditions, and more to provide real-time insights and recommendations. These algorithms can optimize irrigation schedules, predict disease outbreaks, and even customize harvesting based on grape maturity levels. For example, companies like VineView have developed drones equipped with AI-powered imaging systems to monitor vine health and yield predictions. This innovative approach not only increases efficiency and productivity but also enables growers to make more informed decisions that ultimately improve grape quality and wine production. As AI continues to advance, its integration into viticulture is poised to transform the industry by creating a more sustainable and profitable future.

The future of AI in personalized wine recommendations

The future of AI in personalized wine recommendations holds tremendous potential in revolutionizing the way consumers interact with and discover wines. With the advancements in artificial intelligence technology, algorithms can now analyze vast amounts of data including taste preferences, purchase history, and even environmental factors to provide highly tailored wine recommendations. By leveraging machine learning algorithms, these AI systems can continuously learn and adapt to individual preferences, offering more precise and personalized suggestions over time. This approach not only enhances the consumer experience but also provides wineries and retailers with valuable insights into consumer behavior and trends. As the field of AI continues to evolve, the possibilities for personalized wine recommendations are vast, with the potential to shape the future of wine consumption and production in unprecedented ways.

XXV. ROBOTICS AND AUTOMATION IN VITICULTURE

Automation and robotics are revolutionizing the realm of viticulture by offering efficient solutions to labor-intensive tasks in vineyards. Robotic technologies, such as autonomous tractors and drones equipped with advanced sensors, enable precise and timely operations, from soil analysis to crop monitoring. These innovations not only improve productivity but also reduce costs and minimize environmental impact in winemaking. With the integration of artificial intelligence and machine learning algorithms, these systems can optimize vineyard management strategies, leading to higher grape quality and better wine production. Case studies in regions like Napa Valley and Bordeaux have shown significant enhancements in grape yield and wine quality with the implementation of robotic technologies. As the viticulture industry continues to embrace automation, the future holds promising advancements in efficiency, sustainability, and quality in winemaking practices.

The use of robots for vineyard tasks

The use of robots for vineyard tasks marks a significant advancement in the field of oenology and viticulture. By employing robots for tasks such as pruning, harvesting, and monitoring vine health, vineyard management becomes more precise, efficient, and cost-effective. These robots are equipped with sensors and artificial intelligence technology that allow them to navigate vineyards autonomously, collecting data on soil conditions, grape ripeness, and overall vine health. This data can then be analyzed to make informed decisions for optimal grape production. Case studies have shown that the use of robots in vineyards has led to increased yields, improved grape quality, and reduced labor costs. As the wine industry continues to evolve, the integration of robotics into vineyard operations will play a crucial role in shaping the future of oenology and viticulture, offering new opportunities for innovation and growth.

Benefits and limitations of automation

Automation in oenology and viticulture presents a range of benefits and limitations that shape the landscape of wine production. On the positive side, automation can enhance efficiency, accuracy, and consistency in tasks such as harvesting, sorting, and bottling. It can also reduce labor costs, minimize human error, and optimize the use of resources. However, automated systems may lack the nuanced decision-making abilities of experienced human operators, leading to potential limitations in complex processes like grape selection and blending. Additionally, the initial investment in automation technology can be substantial, making it a challenge for smaller wine producers to adopt. Despite these limitations, the benefits of automation in terms of productivity and quality control cannot be overlooked, signaling a significant shift in the way wine is made and setting the stage for continued advances in oenological practices.

Case studies of robotics in action

Case studies of robotics in action highlight the transformative potential of advanced technologies in oenology and viticulture. One such example is the use of autonomous drones equipped with multispectral imaging cameras to monitor vineyard health. These drones can quickly identify areas of stress or disease, allowing for targeted interventions and precision farming practices. In addition, robotic harvesters are revolutionizing grape picking by increasing efficiency and reducing labor costs. The development of robotic pruning systems is another groundbreaking advancement, ensuring consistent and precise trimming of vines. These case studies demonstrate not only the practical applications of robotics in the industry but also the promise they hold for sustainable and innovative practices. By leveraging the power of robotics, oenologists and viticulturists can optimize their processes, enhance quality, and adapt to the ever-evolving challenges of the modern wine industry.

XXVI. WINE AND THE CIRCULAR ECONOMY

The concept of the circular economy is gaining traction in various industries, including agriculture and food production. In the context of oenology, embracing principles of sustainability and waste reduction is crucial for the long-term viability of the industry. By implementing practices such as recycling of grape pomace for agricultural purposes, using renewable energy sources for vineyard operations, and adopting packaging materials that are easily recyclable, wineries can minimize their environmental footprint and contribute to a more sustainable wine production process. Additionally, the circular economy approach can lead to cost savings and increased efficiency, driving innovation and competitiveness in the wine industry. Through the integration of circular economy principles, oenology can not only reduce waste and environmental impact but also pave the way for a more resilient and future-proof industry.

Principles of the circular economy applied to winemaking

Principles of the circular economy can play a significant role in revolutionizing the winemaking industry. By adopting a circular approach, wineries can minimize waste and maximize resource efficiency throughout the entire production process. This involves implementing sustainable practices such as recycling grape pomace to create fertilizers, using wastewater treatment systems to reduce water usage, and repurposing by-products for energy generation. In doing so, wineries can reduce their environmental impact, lower production costs, and create new revenue streams. Embracing the circular economy principles in winemaking not only aligns with consumer preferences for sustainability but also ensures the long-term viability of the industry. This shift towards a more holistic and eco-friendly approach can lead to a more resilient and adaptive wine sector that is better equipped to navigate the challenges of the future.

Upcycling and waste reduction in the wine industry

In the wine industry, upcycling and waste reduction have emerged as essential practices to address environmental concerns and promote sustainability. By upcycling materials such as grape pomace, surplus wine, and used barrels, wineries can transform what would have been waste into valuable products or resources. Through innovative processes like composting grape marc to create organic fertilizer, or repurposing excess wine for vinegar production, wineries are not only reducing waste but also generating additional revenue streams. Furthermore, reusing or refurbishing wine barrels allows for extended use and minimizes the environmental impact of producing new barrels. Embracing upcycling and waste reduction in the wine industry not only aligns with current sustainability goals but also demonstrates a commitment to responsible practices that can shape the future of oenology and viticulture.

Case examples of circular economy practices

Case examples of circular economy practices in the field of oenology and viticulture highlight the innovative approaches being implemented to create a more sustainable industry. One such example is the use of grape marc, a byproduct of winemaking, as a source of bioactive compounds for pharmaceutical and cosmetic applications. This practice not only reduces waste but also generates additional revenue streams for wineries. Another case study involves the implementation of precision agriculture techniques, such as remote sensing and data analytics, to optimize vineyard management practices and minimize resource inputs. By adopting these circular economy practices, wineries can reduce their environmental footprint while enhancing operational efficiency and improving overall product quality. These examples demonstrate the transformative potential of circular economy principles in revolutionizing oenology and viticulture for a more sustainable future.

XXVII. THE IMPACT OF SOCIAL MOVEMENTS ON WINE PRODUCTION

Social movements have played a significant role in shaping the landscape of wine production, influencing everything from grape growing practices to consumer preferences. One such movement that has gained traction in recent years is the push for sustainable and organic winemaking practices. This movement has led to a greater emphasis on environmentally-friendly techniques, such as using natural fertilizers and reducing water consumption. Additionally, social movements advocating for fair labor practices have brought attention to the treatment of vineyard workers and the importance of ethical sourcing in the wine industry. These movements have not only influenced the way wine is produced but have also had a profound impact on consumer behavior, with more and more individuals seeking out wines that align with their values. As social movements continue to evolve, they will undoubtedly play a pivotal role in shaping the future of wine production, guiding the industry towards a more sustainable and socially responsible future.

Fair trade and ethical sourcing in viticulture

In the realm of viticulture, fair trade and ethical sourcing have emerged as essential components of sustainable and responsible wine production. By engaging in fair trade practices, vineyards prioritize equitable pay for workers and environmentally conscious cultivation methods, ensuring a harmonious relationship between labor, land, and community. Ethical sourcing goes beyond mere compliance with regulations, emphasizing transparency, accountability, and social impact. This approach not only benefits the workers and local economies but also resonates with consumers who increasingly seek ethical products. In the context of viticulture, fair trade and ethical sourcing are crucial for promoting the well-being of workers, protecting the environment, and enhancing the overall quality and reputation of wines. As the industry evolves, embracing these principles will be instrumental in fostering a sustainable future for oenology.

The influence of environmental movements on winemaking practices

Environmental movements have had a profound influence on winemaking practices, driving the industry towards more sustainable and eco-conscious methods. Recognizing the importance of preserving the environment and reducing their carbon footprint, winemakers have started to implement innovative techniques such as organic and biodynamic farming, water conservation measures, and biodiversity conservation efforts in their vineyards. These practices not only benefit the environment but also result in higher quality wines with unique terroir expressions. By prioritizing environmental sustainability, winemakers are not only meeting the demands of modern consumers who seek ethically produced products but also laying the foundation for a more resilient and environmentally-friendly industry. As environmental movements continue to gain momentum, it is clear that their influence on winemaking practices will only grow stronger, shaping the future of oenology and viticulture towards a more sustainable and environmentally conscious direction.

Social responsibility and corporate governance in the wine industry

In the wine industry, social responsibility and corporate governance are becoming increasingly pivotal aspects that can significantly impact the overall success and sustainability of businesses. Ethical practices, transparency, and accountability are now recognized as crucial elements in maintaining a positive reputation and building consumer trust. Companies that prioritize social responsibility by engaging in environmentally friendly practices, supporting local communities, and ensuring fair labor conditions not only enhance their brand image but also contribute to the well-being of society as a whole. Moreover, strong corporate governance practices, such as establishing clear decision-making processes, promoting diversity in leadership, and maintaining financial integrity, are essential for fostering long-term growth and mitigating risks. By integrating social responsibility and corporate governance principles into their operations, wine companies can demonstrate their commitment to ethical standards and ultimately drive positive change within the industry.

XXVIII. THE SCIENCE OF WINE FLAVORS AND AROMAS

The study of wine flavors and aromas has undergone a significant transformation in the realm of oenology and viticulture. Advancements in analytical techniques, such as gas chromatography and mass spectrometry, have allowed researchers to dissect the complex profile of wine compounds at a molecular level. By identifying key molecules responsible for specific flavors and aromas, scientists can now manipulate these compounds to enhance desirable characteristics in wines. Additionally, the application of sensory analysis methods, including descriptive analysis and aroma profiling, has enabled a more precise understanding of how different grape varieties, fermentation processes, and aging conditions contribute to the overall sensory experience of wine. Through these scientific advancements, oenologists and viticulturists can now produce wines with unparalleled complexity and quality, ushering in a new era of innovation and excellence in the world of wine production.

Understanding the molecular composition of wine flavors

Understanding the molecular composition of wine flavors is a crucial aspect of oenology that has seen significant advancements in recent years. Through the use of cutting-edge analytical techniques such as gas chromatography-mass spectrometry and nuclear magnetic resonance spectroscopy, researchers have been able to identify and quantify the specific compounds responsible for the sensory characteristics of different wines. By studying the volatile organic compounds, phenolic compounds, and amino acids present in wines, scientists can gain a deeper insight into the complex interactions that give rise to unique flavor profiles. This knowledge not only enhances our appreciation of the nuances in wine taste but also provides valuable information for winemakers seeking to optimize their production processes and create wines of exceptional quality. As oenology continues to evolve, a deeper understanding of wine flavors at the molecular level will undoubtedly play a crucial role in shaping the future of this field.

The role of terpenes and esters in aroma profiles

Terpenes and esters play a crucial role in shaping the aroma profiles of wines, impacting their sensory attributes and overall quality. Terpenes, which are responsible for the floral and fruity notes in wine, are found in grape skins and can be influenced by factors such as grape variety and winemaking techniques. Esters, on the other hand, contribute to the complexity and richness of aromas, forming during fermentation through the interaction of acids and alcohols. Understanding the relationship between terpenes, esters, and aroma profiles allows winemakers to manipulate these compounds to create distinctive and desirable characteristics in their wines. By harnessing the potential of terpenes and esters in aroma development, oenologists can elevate the sensory experience of wine, enhancing its market appeal and establishing a unique identity in the competitive wine industry.

Advances in flavor enhancement and manipulation

Advances in flavor enhancement and manipulation represent a critical aspect of the innovative era in oenology and viticulture. Through cutting-edge techniques, technologies, and methodologies, researchers and winemakers are pushing boundaries to create new flavor profiles that captivate the senses. By utilizing advanced fermentation processes, such as controlled temperature and yeast selection, winemakers can manipulate the aroma and taste components of wine to achieve unique and complex bouquets. For instance, studies have shown that the use of specific enzymes during maceration can enhance the extraction of flavor compounds from grape skins, leading to richer and more intense flavors. These advancements have not only expanded the possibilities for winemaking but have also sparked a renaissance in wine appreciation and exploration. As the field continues to evolve, the potential for flavor manipulation remains a driving force behind the future of oenology and viticulture, opening up exciting new avenues for research and experimentation.

XXIX. THE FUTURE OF WINE CRITICISM AND REVIEWS

As the landscape of oenology and viticulture continues to evolve rapidly, the future of wine criticism and reviews is also poised for significant transformation. Traditional methods of wine evaluation are being challenged by emerging technologies and data-driven approaches that offer more objective and precise assessments. This shift towards a more scientific and analytical approach has the potential to revolutionize the way wines are reviewed and rated, enhancing the reliability and consistency of critiques. With the advent of AI algorithms and sensory analysis tools, reviewers are equipped with new tools to evaluate wines based on a wider range of parameters beyond just taste and aroma. This evolution in wine criticism not only raises the bar for industry standards but also opens up new avenues for exploring the complexities of wine characteristics and production processes, ultimately shaping a future where critiques are more nuanced and insightful.

The changing landscape of wine journalism

In the realm of wine journalism, there has been a notable evolution reflecting the changing landscape of oenology and viticulture. Traditional forms of wine reporting are increasingly complemented by a new wave of digital platforms, social media influencers, and online publications that cater to diverse audiences. This shift has democratized access to wine knowledge and has allowed for a more interactive and engaging communication between producers, consumers, and experts. The emergence of multimedia formats, podcasts, and video content has expanded the scope of wine journalism, offering in-depth insights, virtual tastings, and behind-the-scenes experiences. As a result, the boundaries between journalism, marketing, and education in the wine industry are becoming blurred, challenging traditional notions of authority and expertise. This transformation in wine journalism not only reflects the broader changes in the field but also presents new opportunities and challenges for disseminating knowledge, shaping perceptions, and influencing trends in oenology and viticulture.

The role of online platforms and influencers

In the contemporary landscape of oenology and viticulture, the role of online platforms and influencers has emerged as a powerful force driving consumer engagement and shaping industry trends. Online platforms such as social media channels, blogs, and websites have provided a platform for wineries, vineyards, and wine influencers to connect with a global audience, share insights, and promote products. Through engaging content, stunning visuals, and interactive experiences, these online channels have the ability to influence consumer preferences, educate audiences about wine regions and grape varieties, and even drive sales. Furthermore, the rise of wine influencers, sommeliers, and critics who have amassed large followings on social media platforms has significantly impacted the way consumers perceive and interact with wine-related content. By leveraging the reach and influence of online platforms and influencers, the oenology and viticulture sector can enhance brand visibility, increase consumer trust, and ultimately drive innovation in the industry.

The impact of technology on wine rating systems

The impact of technology on wine rating systems has been profound, revolutionizing the way wines are evaluated and assessed by experts and consumers alike. With the advent of sophisticated sensory analysis tools, artificial intelligence, and big data analytics, wine rating systems have become more objective, accurate, and efficient. These technological advancements have enabled a more systematic and consistent approach to evaluating wines, reducing the subjectivity and bias that were inherent in traditional rating methods. Additionally, new technologies have allowed for the integration of consumer preferences and feedback into the rating process, creating a more dynamic and responsive system that caters to the evolving tastes of wine enthusiasts. As technology continues to advance, the future of wine rating systems looks promising, with the potential to enhance the quality and diversity of wines available in the market.

XXX. WINE AND CLIMATE ADAPTATION STRATEGIES

One of the crucial aspects that shape the future of oenology and viticulture is the relationship between wine production and climate change. In recent years, the wine industry has faced significant challenges due to shifting climate patterns, affecting the quality and quantity of grape harvests. As a response to these challenges, researchers and wine producers have been exploring various adaptation strategies to mitigate the impact of climate change on vineyards. These strategies include the adoption of new grape varieties that are more resilient to extreme weather conditions, implementing innovative irrigation techniques to conserve water in arid regions, and utilizing sensor technology to monitor and optimize vineyard management practices. By incorporating these adaptation strategies, the wine industry can adapt to the evolving climate conditions and continue to produce high-quality wines sustainably. This emphasis on climate adaptation marks a significant shift in the traditional practices of oenology and viticulture, highlighting the necessity for innovation in the face of changing environmental realities.

Breeding grapes for climate resilience

Breeding grapes for climate resilience is a crucial aspect of adapting to the changing environmental conditions in vineyard regions worldwide. By focusing on developing grape varieties that can thrive in extreme temperatures, drought conditions, and unpredictable weather patterns, viticulturists can ensure the sustainability and productivity of their vineyards. This process involves the careful selection and breeding of grape cultivars that exhibit traits such as heat and drought tolerance, disease resistance, and early ripening. Through the use of advanced genetic techniques, such as marker-assisted breeding and genome editing, researchers can accelerate the development of climate-resilient grape varieties. By incorporating these innovative approaches into traditional breeding methods, viticulturists can create grape cultivars that are better equipped to withstand the challenges posed by climate change, ultimately ensuring a more resilient and thriving wine industry in the future.

Vineyard management adjustments for extreme weather

Extreme weather events such as heatwaves, droughts, and storms pose significant challenges to vineyard management, requiring adjustments to be made in order to mitigate their impact on grape quality and vine health. In response to these challenges, vineyard managers are implementing various strategies such as modifying irrigation schedules, implementing canopy management techniques to provide shade and protection, and utilizing protective measures such as hail nets and windbreaks. These adjustments are essential for preserving vineyard productivity and ensuring consistency in wine quality despite unpredictable weather patterns. By incorporating climate data and predictive analytics, vineyard managers can proactively plan for extreme weather events and adapt their practices accordingly. As the frequency and intensity of extreme weather events continue to rise, vineyard management must remain adaptive and innovative to ensure the sustainability and resilience of vineyards in the face of climate change.

Long-term strategies for climate change mitigation

Long-term strategies for climate change mitigation in oenology and viticulture are essential in ensuring the sustainability of the industry. One key approach is the adoption of sustainable practices, such as organic farming, integrated pest management, and water-efficient irrigation systems. These strategies not only reduce greenhouse gas emissions but also promote biodiversity and soil health. Additionally, investing in research and development for resilient grape varieties that can withstand extreme weather conditions is crucial for adapting to a changing climate. Collaboration between wineries, research institutions, and government agencies is essential in sharing knowledge and resources to address climate-related challenges. By implementing these long-term strategies, the oenology and viticulture industry can contribute to mitigating the impacts of climate change while ensuring the quality and longevity of wine production for future generations.

XXXI. INNOVATIONS IN GRAPE HARVESTING

Innovations in grape harvesting represent a crucial advancement in modern oenology and viticulture, offering a glimpse into the transformative potential of technology in the field. With the advent of state-of-the-art machinery and automation, grape harvesting has evolved from a labor-intensive process to a streamlined and efficient operation. Case studies have shown that these innovations not only increase productivity but also improve grape quality by ensuring a timely and precise harvest. For example, the use of robotic harvesters equipped with sensors can selectively pick grapes based on ripeness indicators, resulting in better overall wine quality. These advancements not only revolutionize the harvesting process but also open the door to new possibilities in vineyard management and wine production. As we embrace these changes, it becomes clear that technological innovations are shaping the future of oenology and viticulture, ushering in a new era of possibilities and challenges for research and practice.

Mechanized vs. hand-picking: efficiency and quality

In the realm of oenology and viticulture, the debate between mechanized harvesting and hand-picking has long been a point of contention. Historically, hand-picking was favored for its perceived quality control and gentle handling of grapes, believed to result in higher quality wines. However, recent advancements in mechanized technologies have proven that efficiency and quality need not be mutually exclusive. Mechanized harvesting now offers precise control over picking times, ensuring grapes are harvested at optimal ripeness. Additionally, the technology allows for quicker processing, reducing oxidation and preserving the integrity of the grapes. Studies have shown that wines produced from mechanically harvested grapes can exhibit the same quality as those picked by hand. As oenology and viticulture continue to evolve in this revolutionary era, the choice between mechanized and hand-picking methods will require a nuanced understanding of efficiency, quality, and the overall impact on the final product.

The development of selective harvesting machinery

The development of selective harvesting machinery represents a significant advancement in modern viticulture, offering precision and efficiency in grape harvesting processes. These machines utilize advanced technological features such as sensors, cameras, and sorting mechanisms to selectively pick grapes based on ripeness and quality criteria. By targeting specific clusters or even individual grapes, these machines can optimize grape selection, resulting in higher quality wines with enhanced flavors and aromas. Furthermore, selective harvesting machinery reduces the need for manual labor, saving time and labor costs for vineyard operators. Case studies have shown that the implementation of such machinery has led to improvements in grape quality and overall wine production efficiency. As viticulture continues to evolve in this era of innovation, the development of selective harvesting machinery will undoubtedly play a crucial role in shaping the future of oenology and viticulture, driving further advancements and improvements in wine production practices.

The impact of harvest timing on wine characteristics

The impact of harvest timing on wine characteristics is a critical aspect of oenology and viticulture that can significantly influence the quality and flavor profile of the final product. The timing of grape harvest plays a crucial role in determining the sugar levels, acidity, and phenolic compounds present in the grapes, which in turn affect the taste, aroma, and aging potential of the wine. Research has shown that grapes harvested earlier tend to produce wines with higher acidity and lower alcohol content, while grapes harvested later result in wines with greater ripeness, higher sugar levels, and increased body. Winemakers must carefully consider the optimal timing of harvest based on the desired wine style and regional climate conditions to achieve the best possible outcome in terms of flavor, structure, and overall quality. This nuanced understanding of harvest timing exemplifies the precision and artistry of modern oenology, highlighting the complex interplay between nature, technology, and human expertise in the production of fine wines.

XXXII. THE INTERSECTION OF ART AND SCIENCE IN WINEMAKING

In the ever-evolving realm of winemaking, the intersection of art and science plays a pivotal role in shaping the world of oenology and viticulture. This fusion not only encapsulates the traditional craftsmanship and sensory finesse of winemaking but also integrates cutting-edge scientific knowledge and technological advancements to push the boundaries of innovation. By blending the intuitive skills of an artist with the analytical capabilities of a scientist, winemakers are able to create wines that are not only delicious but also reflect a deeper understanding of terroir, grape varieties, and fermentation processes. The synergy between art and science in winemaking has led to the development of novel techniques, such as precision viticulture and micro-oxygenation, that have revolutionized the way wine is produced and perceived. As we stand at the cusp of this new era in oenology, the harmonious balance between artistic expression and scientific rigor continues to propel the industry forward, offering endless possibilities for exploration and growth.

The balance between creativity and technical knowledge

In the dynamic landscape of oenology and viticulture, the balance between creativity and technical knowledge plays a crucial role in driving innovation and pushing boundaries in winemaking. While technical expertise is essential for understanding the science behind grape growing and fermentation processes, creativity is equally important in developing new flavor profiles, experimenting with unconventional techniques, and appealing to evolving consumer tastes. The synergy between technical know-how and creative thinking is evident in the evolution of wine styles, from traditional methods to avant-garde practices that challenge established norms. By harnessing the power of both creativity and technical knowledge, winemakers can create unique and memorable wines that capture the essence of terroir while pushing the boundaries of what is possible in the world of oenology. This delicate balance is the hallmark of the new era of oenology and viticulture, where innovation and tradition coexist harmoniously to shape the future of winemaking.

The role of the winemaker as an artist

In the realm of oenology, the winemaker is more than a mere technician; they are an artist orchestrating a symphony of flavors and aromas in each bottle of wine they craft. Through their skill, intuition, and creativity, winemakers transcend mere fermentation to create unique expressions of terroir and style. Their decisions on when to harvest, which yeast strains to use, and how to age the wine all contribute to the final masterpiece. Like a painter with their palette, the winemaker selects from a myriad of possibilities to create a harmonious blend that captures the essence of the vineyard and the vintage. This intricate balance of science and artistry is what sets exceptional winemakers apart, allowing them to showcase their individuality and vision in each bottle produced. The role of the winemaker as an artist is quintessential to the revolutionary era of oenology, where innovation and creativity merge to shape the future of winemaking.

Collaborations between scientists and winemakers

In the realm of oenology and viticulture, collaborations between scientists and winemakers have become increasingly pivotal in driving innovation and excellence in wine production. By combining the expertise of scientists in areas such as chemistry, biology, and agricultural science with the practical knowledge and experience of winemakers, new frontiers are being explored and boundaries pushed. These collaborations have led to the development of cutting-edge techniques, technologies, and methodologies that have revolutionized traditional practices and elevated the quality of wines produced. For instance, the integration of precision viticulture techniques, such as remote sensing and GIS mapping, has enabled winemakers to optimize vineyard management practices and maximize grape quality. As these partnerships continue to evolve, the potential for further advancements and discoveries in oenology and viticulture is infinite, shaping the future of the industry and paving the way for a new era of excellence.

XXXIII. THE ECONOMICS OF WINE PRODUCTION

The economics of wine production play a crucial role in the success and sustainability of the industry. As the demand for high-quality wines grows, so does the need for efficient and cost-effective production methods. One key aspect is the impact of innovative technologies on reducing production costs and improving overall efficiency. For example, the use of precision viticulture techniques, such as remote sensing and GPS technology, can optimize vineyard management practices, leading to higher yields and better quality grapes. Additionally, advancements in winemaking equipment and processes have helped streamline production processes, reducing labor costs and increasing productivity. These economic factors are vital in ensuring the competitiveness and profitability of wineries in today's market. By understanding and harnessing the economics of wine production, stakeholders can make informed decisions that drive the industry forward towards a more sustainable and profitable future.

Cost analysis of traditional vs. innovative practices

The critical analysis of costs associated with traditional versus innovative practices in oenology and viticulture reveals a complex landscape influenced by numerous factors. Traditional methods often involve manual labor, increased risk of crop loss due to pest and disease susceptibility, and lower efficiency in resource utilization. On the other hand, innovative practices, such as precision viticulture and sustainable farming techniques, may require initial investment in technology and training but can significantly reduce long-term operational costs and environmental impact. Case studies exemplify the cost-effectiveness of innovative approaches, demonstrating improved crop yield, quality, and resilience to climate change. While the upfront costs of transitioning to innovative practices may be higher, the long-term benefits justify the investment, positioning them as viable solutions for sustainable and profitable oenological and viticultural operations in the future.

Investment in technology and its ROI for wineries

Investment in technology plays a pivotal role in the success of wineries, directly influencing their Return on Investment (ROI) and overall competitiveness in the market. By incorporating cutting-edge technologies such as precision viticulture, data analytics, and automation, wineries can optimize grape cultivation, streamline production processes, and enhance product quality. The implementation of these technologies enables wineries to increase efficiency, reduce costs, and improve overall sustainability, ultimately leading to a higher ROI. For example, by utilizing drones for vineyard monitoring or software for predictive analytics, wineries can make informed decisions that positively impact their bottom line. As the wine industry becomes increasingly competitive and consumer preferences evolve, investing in technology is no longer a luxury but a necessity for wineries looking to stay ahead of the curve and thrive in this revolutionary era of oenology and viticulture.

Economic models for sustainable viticulture

Economic models for sustainable viticulture play a critical role in shaping the future of oenology by providing a framework for balancing environmental and economic considerations. These models encompass various factors such as resource management, cost analysis, and market trends to guide decision-making processes in vineyard management. By implementing sustainable practices, vineyard owners can not only reduce their environmental impact but also improve long-term profitability and resilience in the face of changing market conditions. One such model is the concept of agroecology, which emphasizes the integration of ecological principles into agricultural systems to increase biodiversity and promote natural pest control. Through comprehensive economic analysis, including factors such as the cost of inputs, labor, and potential market returns, vineyard owners can make informed decisions that prioritize sustainability without compromising economic viability. As viticulture continues to evolve in this revolutionary era, economic models for sustainable practices will be crucial in ensuring the industry's long-term success.

XXXIV. THE ROLE OF CONSORTIA AND WINE ASSOCIATIONS

In the landscape of modern oenology and viticulture, consortia and wine associations play a crucial role in fostering collaboration, sharing knowledge, and advancing the industry as a whole. These organizations bring together winemakers, researchers, and industry professionals to exchange ideas, develop best practices, and collectively address challenges facing the sector. Through joint research projects, educational initiatives, and marketing campaigns, consortia and wine associations promote innovation and sustainability in grape growing and winemaking. By pooling resources and expertise, members of these groups can access cutting-edge technologies, scientific findings, and market insights that may not be readily available to individual producers. As the wine industry continues to evolve, the collaborative efforts facilitated by consortia and wine associations will be essential in driving progress and shaping the future of oenology and viticulture.

The function of wine consortia in regulation and promotion

Wine consortia play a crucial role in both regulating and promoting the wine industry. These organizations often bring together vineyards, wineries, and other stakeholders within a specific region or appellation to collectively work towards quality standards, sustainability practices, and marketing efforts. By setting guidelines and standards for production, wine consortia help ensure consistency and authenticity in the wines produced within their region. Additionally, these consortia engage in promotional activities such as wine festivals, tastings, and marketing campaigns to raise awareness and increase demand for their wines both domestically and internationally. Through collaboration and collective action, wine consortia are able to elevate the reputation of their region's wines, strengthen consumer confidence, and drive economic growth. As we enter this revolutionary era of oenology and viticulture, the role of wine consortia will continue to be integral in shaping the future of the industry.

Collaborative efforts for research and development

Collaborative efforts for research and development play a pivotal role in driving innovation and progress within the realm of oenology and viticulture. By bringing together experts from various disciplines such as chemistry, biology, and engineering, collaborative projects have the potential to break new ground and push the boundaries of traditional winemaking practices. These endeavors often involve joint research initiatives between academic institutions, industry partners, and government agencies, pooling resources and expertise to tackle complex challenges in the field. Through shared knowledge, data, and resources, researchers can accelerate the development of cutting-edge technologies and techniques that can revolutionize the way wine is produced and enjoyed. The collaborative nature of these efforts not only fosters creativity and innovation but also ensures that the benefits of research and development are accessible to the broader winemaking community, ultimately shaping the future of oenology and viticulture in profound ways.

Case studies of successful consortia initiatives

Examining case studies of successful consortia initiatives reveals the significant impact collaborative efforts can have on advancing oenology and viticulture. One notable example is the consortium formed in the Loire Valley, where winemakers, researchers, and industry experts joined forces to develop sustainable viticultural practices. Through this partnership, innovative techniques such as organic and biodynamic farming have been implemented, leading to improved wine quality and environmental stewardship. In another case study, a consortium in California focused on research and development of new grape varieties resistant to climate change. By leveraging collective expertise and resources, these initiatives have paved the way for novel approaches in winemaking and grape cultivation. These successful consortia initiatives underscore the power of collaboration in driving innovation and shaping the future of the wine industry. As we continue to embrace this new era of oenology, exploring further collaborative endeavors will be vital in addressing future challenges and opportunities.

XXXV. WINE LABELING AND CONSUMER INFORMATION

In the realm of oenology, the labeling and dissemination of consumer information play a crucial role in guiding consumer choices and shaping perceptions of wine products. With the rise of discerning and knowledgeable consumers, there is an increasing demand for transparency and clarity in wine labeling. Labels are not merely decorative but serve as a means of communication between producers and consumers, conveying vital information such as grape variety, origin, vintage, and production methods. As such, innovative approaches to wine labeling, such as QR codes linking to online platforms with detailed information, allergen warnings, and sustainability certifications, have emerged to meet these demands. By providing consumers with comprehensive and accurate information, wine labels can empower them to make informed decisions, enhance trust in the product, and ultimately contribute to a more transparent and sustainable wine industry. This evolution in wine labeling reflects the ongoing revolution in oenology and viticulture towards greater consumer engagement and satisfaction.

The evolution of wine labels as a source of information

The evolution of wine labels as a source of information has been a significant development in the field of oenology and viticulture. Traditionally, wine labels were primarily used for branding and marketing purposes, with limited information about the wine itself. However, in recent years, there has been a shift towards more informative and educational labels. These modern labels often include details about the grape varietals used, the vineyard location, the winemaking process, and even recommended food pairings. This shift can be attributed to a growing consumer interest in understanding the origin and production of the wines they consume. By providing more detailed information on their labels, wineries are able to connect with consumers on a deeper level, enhancing the overall wine-drinking experience. As technology continues to advance, we can expect wine labels to become even more informative and interactive, further enhancing the consumer's knowledge and appreciation of wine.

The role of labels in communicating wine quality and origin

Labels play a crucial role in communicating the quality and origin of wine to consumers. These labels serve as a strategic tool for producers to convey important information about the product, such as the grape variety, region of production, and vintage. By including this data on the label, consumers can make more informed choices when selecting a bottle of wine, leading to increased satisfaction and loyalty. Furthermore, labels can also convey a sense of prestige and exclusivity, influencing consumers' perceptions of the wine's quality and value. In the context of the revolutionary era in oenology and viticulture, labels can be instrumental in showcasing innovative techniques and sustainable practices used in the production process, fostering a deeper connection between consumers and the wine they are enjoying. As advancements in technology continue to reshape the industry, labels will undoubtedly play an even more significant role in shaping consumer perceptions and driving market trends.

Regulatory changes and trends in wine labeling

Regulatory changes and trends in wine labeling play a significant role in shaping the landscape of the wine industry. As consumer awareness and interest in wine production continue to grow, the need for clear and transparent labeling has become paramount. The beginning of this revolution in wine labeling can be seen in the increased emphasis on sustainability, organic practices, and origin labeling. These changes reflect the industry's response to evolving consumer preferences and demands for more information about the products they purchase. In the middle of this transition, we also see advancements in digital labeling technologies, allowing producers to provide detailed information about their wines in a more interactive and engaging manner. As we move towards the end of this transformation, it is clear that regulations will continue to evolve to ensure the integrity and accuracy of wine labeling, setting a new standard for transparency and authenticity in the industry.

XXXVI. THE INFLUENCE OF WINE ON CULTURE AND SOCIETY

Wine has long played a crucial role in shaping cultural practices and social interactions around the world. From ancient rituals to modern celebrations, wine has been an integral part of human society, influencing art, literature, religion, and even politics. The sheer diversity of wines available reflects the rich tapestry of cultures that have embraced this beverage throughout history. By examining the impact of wine on cultural norms and societal structures, we can gain a deeper understanding of the connections between human behavior and the consumption of alcohol. As we delve into this intricate relationship, it becomes apparent that wine not only serves as a means of social lubrication but also as a vehicle for expressing identity, status, and traditions. Through a comprehensive study of the influence of wine on culture and society, we can unravel the complex web of interactions that have shaped our collective history and continue to impact our contemporary world.

Wine as a cultural symbol and its historical significance

Wine has been a cultural symbol with historical significance for centuries, transcending geographical and societal boundaries. In ancient societies, the production and consumption of wine were intertwined with religious rituals, social gatherings, and political ceremonies, shaping the identities and beliefs of civilizations. As a cultural symbol, wine has evolved to represent luxury, sophistication, tradition, and heritage, with each region producing its unique varietals that reflect its terroir and winemaking practices. The historical significance of wine can be seen in its influence on art, literature, and cuisine, showcasing its enduring impact on human civilization. Today, the cultural symbolism of wine continues to play a vital role in societal customs, celebrations, and even diplomatic relations, highlighting its timeless appeal and relevance in shaping human interactions and experiences.

The social dynamics of wine consumption

The social dynamics of wine consumption play a crucial role in the success and popularity of oenology and viticulture. Wine has long been associated with social gatherings, celebrations, and rituals, creating a sense of communal experience and shared enjoyment. The choice of wine, the manner in which it is served, and the conversations it inspires all contribute to the intricate social fabric surrounding wine. Additionally, trends in wine consumption reflect broader cultural shifts, with preferences for certain varieties or styles often mirroring societal values and preferences. Understanding the social dynamics of wine consumption is essential for producers and marketers in this field, as it can inform strategies for product development, branding, and promotion. By recognizing the significance of social factors in wine consumption, researchers and practitioners can continue to innovate and adapt to changing consumer behaviors and preferences in this dynamic industry.

Wine's role in contemporary cultural practices

In contemporary cultural practices, wine plays a pivotal role as both a symbol of luxury and a means of socialization. The consumption and appreciation of wine have become integral to social gatherings, fine dining experiences, and celebrations, thereby contributing to the cultivation of sophisticated tastes and refinement in society. Wine tastings, wine tours, and wine pairings have all become popular activities in which individuals engage to enhance their understanding and enjoyment of this ancient beverage. Furthermore, the cultivation of specific grape varieties and the production of unique blends have allowed winemakers to showcase their creativity and expertise, resulting in a diverse array of wines that cater to a wide range of preferences. The cultural significance of wine continues to evolve, influencing not only consumer behavior but also societal norms and values as oenology and viticulture embrace innovation in the quest for excellence.

XXXVII. THE PSYCHOLOGY OF WINE TASTING

The Psychology of Wine Tasting is a fascinating area of study that delves into the intricate relationship between our senses, cognition, and emotions when evaluating the complex flavors and aromas of wine. Through sensory analysis techniques and experimental methodologies, researchers have been able to uncover how factors such as context, expectations, and personal preferences can greatly influence our perception of wine. This understanding has profound implications for the wine industry, as it reveals the importance of creating optimal tasting environments and experiences that enhance consumer enjoyment and appreciation. By integrating insights from psychology into traditional oenological practices, winemakers can refine their techniques and refine their products to better align with consumer preferences and trends. The Psychology of Wine Tasting not only enriches our understanding of sensory perception but also offers valuable insights for improving wine production and marketing strategies in the modern era of oenology.

Cognitive aspects of taste and aroma perception

One of the key cognitive aspects of taste and aroma perception in oenology is the concept of sensory evaluation. This plays a crucial role in understanding how individuals perceive and interpret the complex flavors and aromas present in wines. By examining the impact of factors such as experience, expectations, and cultural background on sensory perception, researchers can gain insight into how these elements influence the overall tasting experience. Recent studies have shown that cognitive processes, such as memory and attention, play a significant role in shaping individuals' perceptions of taste and aroma. These findings highlight the importance of considering not only the chemical composition of wines but also the cognitive processes involved in sensory evaluation. By incorporating cognitive aspects into the study of taste and aroma perception, oenologists can gain a deeper understanding of how these elements interact to create the sensory experience of wine.

The influence of context and expectations on wine tasting

The influence of context and expectations on wine tasting is a multifaceted and often overlooked aspect of oenology. Research has shown that environmental factors, such as lighting, background noise, and even the appearance of the bottle, can significantly affect how a wine is perceived by tasters. Moreover, the expectations that individuals bring to a tasting can also play a powerful role in shaping their experience. Studies have demonstrated that people who believe they are drinking an expensive wine tend to rate it more favorably, even if it is actually a lower-priced bottle. This suggests that our perception of wine is not solely based on its intrinsic qualities, but is heavily influenced by our surroundings and mental predispositions. Understanding and accounting for these contextual factors in wine tasting can lead to a more nuanced appreciation of the complexities of this ancient art form.

Experimental studies on the psychology of wine preferences

Experimental studies on the psychology of wine preferences play a crucial role in understanding consumer behavior and guiding marketing strategies within the wine industry. Through controlled experiments and sensory analysis, researchers have been able to delve into the intricate factors that influence an individual's wine choices. Studies have shown that elements such as labeling, pricing, and even the shape of the wine glass can significantly impact one's perception and enjoyment of a wine. By employing methods such as blind tastings and psychological surveys, researchers can uncover hidden biases and preferences that shape consumer decisions. These experimental studies not only provide valuable insights for winemakers and marketers but also contribute to the broader field of sensory psychology. As these studies continue to evolve, the findings will undoubtedly shape the future landscape of wine production and marketing strategies, emphasizing the importance of understanding the intricate interplay between psychology and wine preferences.

XXXVIII. THE INTEGRATION OF WINE IN CULINARY ARTS

The integration of wine in culinary arts has been a longstanding tradition that continues to evolve and thrive in the modern era of gastronomy. Wine not only acts as a complimentary beverage to various dishes but also serves as an essential ingredient in recipes, adding depth, complexity, and a unique flavor profile to the final dish. Chefs and sommeliers are now collaborating more closely than ever, exploring innovative pairings and techniques to enhance the overall dining experience. By understanding the intricate relationship between wine and food, professionals in both the wine and culinary industries are revolutionizing traditional practices and creating new gastronomic trends. This integration not only elevates the dining experience for consumers but also pushes the boundaries of creativity and experimentation in both fields, shaping the future of oenology and culinary arts. As this trend continues to gain momentum, the possibilities for culinary innovation and wine appreciation are endless.

Wine pairing principles and contemporary trends

As the landscape of oenology and viticulture continues to evolve, the principles of wine pairing have also undergone a transformation. Traditional guidelines based on regional origins and grape varietals have given way to more innovative and dynamic approaches. Contemporary trends in wine pairing now focus on a deeper understanding of flavor profiles, textures, and aromas to create harmonious combinations. Techniques such as molecular gastronomy and sensory analysis are being integrated into the process, allowing for more precise and personalized pairings. Additionally, the rise of organic and biodynamic wines has brought about a shift towards sustainability and natural practices in wine production, influencing pairing choices. Moving forward, it is essential for sommeliers and wine enthusiasts to stay abreast of these trends to enhance the dining experience and elevate the appreciation of wine.

The collaboration between chefs and winemakers

Collaboration between chefs and winemakers is a dynamic partnership that has evolved significantly in recent years, contributing to the enhancement of the overall dining experience. By working closely together, chefs and winemakers can create harmonious pairings that elevate the flavors of both food and wine. Through experimentation and innovation, these collaborations have led to the development of new culinary techniques and wine styles that push the boundaries of traditional pairings. For instance, the use of molecular gastronomy techniques in food preparation has inspired winemakers to craft wines with unique flavor profiles that complement these avant-garde dishes. This trend towards collaboration highlights the importance of cross-disciplinary partnerships in the culinary and viticultural worlds, indicating a shift towards a more creative and forward-thinking approach to food and wine pairings. As this collaboration continues to evolve, it has the potential to revolutionize the way we think about dining and drinking, setting new standards for excellence in gastronomy.

The role of wine in gastronomy and fine dining

In the realm of gastronomy and fine dining, wine plays a pivotal role in enhancing the dining experience and complementing the flavors of a dish. The intricate relationship between wine and food has long been studied and celebrated, with oenophiles and chefs alike recognizing the importance of selecting the perfect wine to elevate a meal. The sensory experience of wine tasting, with its intricate aromas, flavors, and textures, adds a layer of complexity to the dining experience, creating a harmonious balance between the dish and the beverage. In fine dining establishments, sommeliers play a crucial role in guiding patrons through the extensive wine list, highlighting the nuances of different varietals and regions. As the culinary world continues to evolve, the role of wine in gastronomy will undoubtedly remain a cornerstone of the dining experience, enriching the palate and engaging all the senses.

XXXIX. WINE AND THE GIG ECONOMY

Wine and the Gig Economy encapsulate a fascinating intersection between traditional viticulture practices and modern consumer demands. In this era of rapid technological advancement and changing consumer preferences, the wine industry is witnessing a shift towards flexible employment arrangements and personalized wine experiences. The Gig Economy, characterized by short-term contracts and freelance work, is reshaping how wineries market and sell their products, with a focus on direct-to-consumer models and customized wine subscriptions. This novel approach not only provides wineries with new avenues for reaching consumers but also offers wine enthusiasts the opportunity to access unique and tailored wine experiences. By leveraging technology and embracing the principles of the Gig Economy, wineries can create innovative platforms that cater to individual preferences and enhance customer engagement, ultimately driving growth and sustainability in the dynamic world of oenology and viticulture.

The rise of freelance work in the wine industry

The rise of freelance work in the wine industry marks a significant shift in the traditional employment structures within oenology and viticulture. With the increasing demand for specialized skills and expertise in grape cultivation, winemaking, and vineyard management, freelance professionals are now able to offer their services on a project basis, providing flexibility and diversity to wineries and vineyards. This trend allows for the integration of cutting-edge research, innovative techniques, and emerging technologies in wine production, as freelancers can bring fresh perspectives and tailored solutions to specific challenges. By collaborating with freelance experts, wineries can gain access to a broader talent pool and accelerate their innovation processes. However, this shift also presents challenges in terms of project management, coordination, and quality control, prompting the need for effective communication and collaboration strategies to ensure the success of freelance-based projects in the wine industry.

The impact of the gig economy on traditional wine jobs

The gig economy, characterized by temporary and freelance work arrangements facilitated by digital platforms, has been making waves across various industries, including the wine sector. Traditional wine jobs, such as vineyard workers and tasting room staff, are experiencing a shift as more wineries turn to gig workers for flexibility and cost-effectiveness. This shift has implications for job stability, benefits, and quality of work as gig workers may lack the same protections and opportunities for advancement as traditional employees. Moreover, the gig economy can lead to greater competition among workers and a potential decrease in job security in the industry. As oenology and viticulture continue to evolve in this revolutionary era, it is essential to consider the impact of these changes on the labor force and the overall sustainability of the wine industry.

Opportunities and challenges for gig workers in oenology

Opportunities and challenges for gig workers in oenology present a dual-sided aspect to the rapidly evolving field. On one hand, the gig economy offers freelance oenologists flexibility in choosing projects and schedules, enabling them to gain diverse experiences and build a versatile skill set. This allows for greater innovation and creativity in the industry, as gig workers bring fresh perspectives and varied expertise to different projects. However, challenges such as job insecurity, lack of benefits, and difficulty in establishing long-term relationships with clients can hinder the growth and stability of gig workers in oenology. To address these issues, further research is needed to explore ways to provide gig workers with more support, such as access to training programs, networking opportunities, and avenues for professional development. By striking a balance between opportunities and challenges, the gig economy in oenology has the potential to revolutionize the field while also ensuring the well-being and sustainability of its workers.

XL. THE ROLE OF WINE IN SUSTAINABLE DEVELOPMENT GOALS

The sustainable development goals (SDGs) provide a comprehensive framework for addressing global challenges, including those related to environmental sustainability and social equity. Wine production, as a significant agricultural industry, plays a pivotal role in achieving these goals. Through the adoption of sustainable practices in viticulture and oenology, the wine industry can contribute to environmental conservation, resource efficiency, and community empowerment. Practices such as organic and biodynamic farming, water management strategies, and energy efficiency measures not only reduce the environmental footprint of wine production but also promote biodiversity and soil health. Furthermore, initiatives that support fair labor practices and local communities can enhance social well-being and economic development. By aligning with the SDGs, the wine industry can not only improve its sustainability credentials but also contribute to a more equitable and resilient future for all stakeholders involved.

Aligning wine production with the UN's SDGs

As the wine industry embraces a new era of innovation, aligning wine production with the United Nations' Sustainable Development Goals (SDGs) has become imperative. By integrating sustainable practices into viticulture and oenology, wineries can contribute to global efforts towards environmental and social responsibility. Techniques such as organic and biodynamic farming, as well as water and energy-efficient practices, can reduce the carbon footprint of wine production and protect natural resources. Additionally, supporting local communities and promoting fair labor practices align with the SDGs' focus on inclusive economic growth and social equity. Case studies highlighting successful implementation of these strategies demonstrate the viability and benefits of aligning wine production with the SDGs. Moving forward, it is essential for the wine industry to continue prioritizing sustainability and ethical practices to ensure a more prosperous and environmentally conscious future.

Case studies of wineries contributing to sustainable development

Case studies of wineries contributing to sustainable development provide valuable insights into the way in which the wine industry is evolving in response to environmental concerns and social responsibility. By examining the practices and strategies adopted by these wineries, we can gain a better understanding of how sustainable development can be effectively implemented in the viticulture and oenology sectors. Through the adoption of sustainable practices such as organic farming, water conservation, and energy efficiency, wineries are not only reducing their environmental impact but also improving their overall efficiency and competitiveness in the market. These case studies highlight the importance of a holistic approach to sustainability, encompassing not only environmental considerations but also social and economic aspects. By learning from these successful examples, other wineries can also strive towards becoming more sustainable, ultimately contributing to a more environmentally friendly and socially responsible wine industry.

The potential for wine to drive social and environmental progress

The potential for wine to drive social and environmental progress is a topic that has garnered increasing attention in recent years within the field of oenology and viticulture. As wine production continues to expand globally, the industry faces growing pressure to adopt more sustainable and socially responsible practices. By embracing initiatives such as organic and biodynamic farming, water and energy conservation, and fair labor practices, winemakers have the opportunity to not only minimize their environmental impact but also positively influence the communities in which they operate. Furthermore, with the rise of certifications such as "B Corp" and "Sustainable Winegrowing," consumers are becoming more conscious of the social and environmental implications of their purchasing decisions. This shift towards sustainability in the wine industry has the potential to drive significant progress towards a more environmentally conscious and socially responsible future.

XLI. THE IMPACT OF EXCHANGE RATES AND TARIFFS

In the realm of oenology and viticulture, the impact of exchange rates and tariffs cannot be overlooked. The fluctuation of exchange rates can significantly influence the cost of importing or exporting wine, affecting the profitability of wineries and the affordability of wines for consumers. Similarly, tariffs imposed on wine imports can create barriers to entry for foreign wineries, potentially limiting the variety and availability of wines in the market. Understanding the dynamics of exchange rates and tariffs is crucial for industry players to navigate the global wine trade successfully. By analyzing these economic factors and their implications on the wine industry, researchers can gain insights into strategies for mitigating risks and maximizing opportunities in a rapidly changing market. As the landscape of oenology and viticulture evolves, addressing the challenges posed by exchange rates and tariffs will be essential for shaping a sustainable future for the industry.

The influence of currency fluctuations on wine trade

The influence of currency fluctuations on wine trade is a significant factor that impacts the global market for wine. Currency fluctuations can affect the cost of production, distribution, and sales of wine, ultimately impacting the competitiveness of wine producers in different regions. As currencies fluctuate, the price of wine can vary, making it more expensive for consumers in some countries and potentially leading to changes in consumption patterns. Wine producers must carefully monitor currency movements and adjust their strategies to mitigate the risks associated with currency fluctuation. Factors such as exchange rates, inflation rates, and geopolitical events can all contribute to currency fluctuations, making it essential for wine producers to stay informed and adaptable in their trade practices. By understanding and effectively managing the impact of currency fluctuations, wine producers can navigate the complexities of the global market and maintain a competitive edge in the industry.

The effects of tariffs on international wine markets

The effects of tariffs on international wine markets are crucial to understanding the dynamics of global trade in the wine industry. Tariffs can significantly impact the competitiveness of wine-producing countries and influence consumer choices. When tariffs are imposed on imported wines, it can lead to higher prices for consumers in the importing countries, potentially reducing demand for foreign wines. This, in turn, affects the export revenues of wine-producing countries, particularly those that heavily rely on international markets. However, tariffs can also be used as a strategy to protect domestic wine producers by making imported wines less competitive. Understanding the implications of tariffs on international wine markets is essential for policymakers and industry stakeholders to make informed decisions that promote fair trade practices and sustainable growth in the global wine industry.

Strategies for wineries to mitigate economic risks

In the realm of wineries, economic risks are a primary concern that can significantly impact the sustainability and profitability of operations. To mitigate these risks effectively, wineries can adopt several strategic measures. Firstly, diversification of product offerings can help reduce dependence on a single revenue stream, providing a buffer against market fluctuations. Additionally, establishing strong relationships with distributors and retail partners can create stable sales channels and foster long-term partnerships. Moreover, implementing efficient cost management practices, such as optimizing production processes and monitoring expenses closely, can enhance overall financial resilience. Embracing technological advancements, such as precision viticulture and data analytics, can also enable wineries to make more informed decisions and improve resource allocation. By combining these strategies, wineries can proactively manage economic risks and position themselves for long-term success in a competitive market landscape.

XLII. THE ROLE OF WINE IN INTERNATIONAL DIPLOMACY

Wine has played a significant role in international diplomacy throughout history, serving as a symbol of hospitality, culture, and social bonding. The exchange of wine between nations has often been used as a diplomatic tool to foster relationships, facilitate negotiations, and build alliances. From ancient times to the modern era, the sharing of wine has been a common practice during diplomatic meetings, state banquets, and official ceremonies. This practice not only highlights the cultural significance of wine but also demonstrates the role it plays in bringing people together and promoting goodwill between nations. In the context of modern international relations, the strategic use of wine as a diplomatic gift or gesture continues to hold relevance, showcasing the enduring influence of this age-old tradition in shaping diplomatic interactions on a global scale.

Wine as a tool for cultural exchange and diplomacy

One noteworthy aspect of the revolutionary era in oenology and viticulture is the role of wine as a tool for cultural exchange and diplomacy. Wine has a long history of bringing people together, transcending borders and fostering communication between nations. Through the sharing and appreciation of different varietals and winemaking traditions, individuals can connect on a deeper level, bridging cultural gaps and promoting understanding. In recent years, wine diplomacy has gained traction as a strategy for building relationships and resolving conflicts in a non-confrontational manner. By utilizing wine as a common ground, diplomats can engage in dialogue and negotiation in a more relaxed and congenial atmosphere. This approach not only enhances diplomatic efforts but also promotes cultural diversity and mutual respect, reinforcing the idea that wine can be more than just a beverage - it can be a powerful instrument for promoting peace and unity on a global scale.

Historical instances of wine influencing diplomatic relations

Historical instances of wine influencing diplomatic relations have been well-documented throughout the annals of time, highlighting the significant role that wine plays in fostering international relations and negotiations. The ancient Greeks and Romans utilized wine as a tool for diplomacy, sharing and exchanging it as a symbol of goodwill and unity. Fast forward to the modern era, and we see countries using wine as a diplomatic gift to strengthen ties and facilitate discussions. For example, in recent years, leaders have exchanged rare and exquisite bottles of wine to signify the importance of their diplomatic relationship. The symbolic nature of wine as a gift transcends borders and cultures, emphasizing the shared appreciation of this ancient beverage and its ability to bring people together in negotiation and collaboration. As we delve further into the new era of oenology and viticulture, it is essential to acknowledge the historical foundations of wine diplomacy and its enduring influence on diplomatic relations.

The potential for wine to bridge international divides

Wine has the potential to transcend borders and bring people together in a shared appreciation for its complexities and nuances. As a culturally significant beverage with a rich history, wine can serve as a common ground for individuals from diverse backgrounds to connect and engage in meaningful dialogue. By fostering an environment of conviviality and camaraderie, wine has the power to bridge international divides and promote understanding among disparate communities. Through wine tastings, vineyard tours, and educational seminars, individuals can explore new perspectives and gain a deeper appreciation for different cultures and traditions. As the global wine industry continues to expand and evolve, the potential for wine to serve as a unifying force in an increasingly interconnected world is immense. By recognizing and embracing the shared humanity in each bottle, wine enthusiasts can pave the way for a more harmonious and inclusive future.

XLIII. WINE AUTHENTICATION AND FRAUD PREVENTION

In today's wine market, where authenticity is paramount, the issue of wine fraud is becoming increasingly prevalent. Wine authentication and fraud prevention have emerged as crucial areas in the field of oenology, with the necessity to ensure that consumers are getting what they pay for. Technological advancements, such as DNA testing, isotopic analysis, and blockchain technology, have revolutionized the way wines are authenticated and tracked throughout the supply chain. By using these tools, researchers and industry professionals can verify the origin and authenticity of wines, ultimately safeguarding against counterfeiting and fraudulent practices. Case studies and data analysis have shown the effectiveness of these methods in detecting and preventing wine fraud, highlighting the significance of incorporating these techniques into the standard procedures of the wine industry. Moving forward, a continued emphasis on innovative technologies and stringent authentication processes will be vital in maintaining the integrity of the wine market and ensuring consumer trust.

Technologies for ensuring wine authenticity

One of the critical aspects of wine production is ensuring its authenticity, which has become increasingly important in a market plagued by counterfeiting and adulteration. To address this issue, advanced technologies have emerged to guarantee the integrity of wines. For instance, DNA analysis is being utilized to authenticate grape varieties and detect any genetic modifications. Isotope analysis, on the other hand, can determine geographical origin and verify labels claiming specific regions. Additionally, the use of blockchain technology is revolutionizing the supply chain, providing a transparent and immutable record of each wine's journey from vineyard to consumer. These technologies not only protect the reputation of winemakers but also safeguard consumers from fraudulent practices. As the wine industry enters a new era of oenology and viticulture, these innovative measures are crucial in upholding the authenticity and quality of wines in an ever-evolving market.

The global impact of wine fraud

Wine fraud has a global impact on the wine industry, affecting not only consumers but also producers, distributors, and regulators. The prevalence of counterfeit wines, mislabeling, and adulteration has negative consequences for the industry's reputation and economic stability. The consequences extend beyond financial losses to encompass health risks and legal implications. The rise of online marketplaces has made it easier for fraudulent activities to go undetected, posing a significant challenge for enforcement agencies. Collaboration between international organizations, governments, and industry stakeholders is essential to combatting wine fraud effectively. Implementing stricter regulations, enhancing traceability measures, and investing in advanced authentication technologies are crucial steps to safeguarding the integrity of the global wine market. Addressing these issues will not only protect consumers but also uphold the authenticity and quality standards of the industry, ensuring its long-term sustainability and growth.

Legal and regulatory measures to combat counterfeit wines

Counterfeit wines pose a significant threat to the integrity and reputation of the wine industry, leading to economic losses and potential health risks for consumers. In response, legal and regulatory measures have been implemented to combat this issue. These measures include strict labeling requirements, certification programs, and enforcement actions to prosecute counterfeiters. For instance, the European Union has established a geographical indication system to protect the names of wines from specific regions, ensuring consumers receive authentic products. Additionally, customs agencies around the world work to identify and seize counterfeit wines entering their countries. While these measures are essential steps in safeguarding the industry, ongoing collaboration between government agencies, industry stakeholders, and international organizations is crucial to effectively combating counterfeit wines and maintaining consumer trust in the market. Ultimately, robust legal and regulatory frameworks are essential in preserving the authenticity and quality of wines in the global market.

XLIV. THE SCIENCE OF WINE PRESERVATION

In the realm of oenology, the preservation of wine is a critical aspect that directly impacts the quality and longevity of the final product. With the advances in technology and research, there have been significant innovations in the science of wine preservation. Techniques such as inert gas preservation, vacuum sealing, and temperature control have revolutionized the way winemakers can preserve their wines, maintaining their flavor and aroma profiles over extended periods. These methods have been shown to effectively reduce oxidation and microbial contamination, preserving the integrity of the wine. Case studies have demonstrated the effectiveness of these preservation techniques in high-end wineries, showcasing their potential to enhance the quality and shelf life of wines. As the oenological landscape continues to evolve, the science of wine preservation plays a crucial role in ensuring that the art of winemaking can be enjoyed for generations to come.

Advances in wine storage technology

Advances in wine storage technology have significantly transformed the landscape of oenology and viticulture. With the introduction of temperature-controlled fermentation tanks, the process of winemaking has become more precise, allowing for the preservation of delicate flavors and aromas. Additionally, advancements in barrel technology, such as the use of oak alternatives and micro-oxygenation systems, have revolutionized the aging process, enabling winemakers to achieve desired flavor profiles in a shorter amount of time. The advent of inert gas systems has also played a crucial role in preventing oxidation and ensuring the longevity of wines. These innovations not only enhance the quality of wines but also offer greater flexibility and efficiency in production. As technology continues to advance, the future of wine storage holds immense potential for further improvement and refinement, driving the evolution of the industry towards new horizons.

The chemistry of wine spoilage and preservation

The chemistry of wine spoilage and preservation is a critical aspect of oenology that directly impacts the quality and longevity of wine. Spoilage can occur due to various factors, including microbial contamination, oxidation, and chemical reactions. Microbial spoilage, often caused by bacteria, yeast, or mold, can result in off-flavors, cloudiness, and even complete wine ruin. Oxidation, on the other hand, can lead to the loss of fruity aromas and browning of wine. Preserving wine involves understanding and controlling these chemical reactions through techniques such as sulfur dioxide addition, filtration, and temperature control. Advanced analytical tools, such as gas chromatography-mass spectrometry, have enabled researchers to detect and quantify spoilage compounds accurately. By studying the chemistry of wine spoilage and preservation, oenologists can develop effective strategies to maintain the quality and authenticity of wine, ensuring consumer satisfaction and industry sustainability.

Innovations in extending wine shelf life

Innovations in extending wine shelf life have become a critical focus in the realm of oenology and viticulture, allowing producers to maintain the quality and freshness of their products for prolonged periods. Advanced techniques such as micro-oxygenation and membrane filtration have emerged as groundbreaking tools in enhancing the stability and aging potential of wines. By carefully controlling oxygen levels during the winemaking process, micro-oxygenation can mitigate oxidation, improve color stability, and enhance the overall structure of the wine. Similarly, membrane filtration offers a precise method to remove unwanted particulates and microbes, thus ensuring microbiological stability and clarity in the final product. These innovations not only extend the shelf life of wines but also contribute to maintaining their sensory characteristics and quality over time. As the industry continues to embrace these advancements, the future of oenology and viticulture appears promising, with new possibilities for experimentation and refinement on the horizon.

XLV. THE ROLE OF WINE IN FESTIVALS AND EVENTS

The presence of wine has long been intertwined with cultural celebrations and events, playing a significant role in enhancing the overall experience. Festivals and gatherings often feature wine as a central element, serving as a symbol of conviviality, tradition, and indulgence. Wine tastings, wine pairings, and wine auctions are popular components of various festivals, allowing participants to explore different varieties, regions, and styles. Moreover, winemakers often showcase their finest creations at these events, highlighting the craftsmanship and artistry behind each bottle. The sensory delight of savoring a quality wine in a festive atmosphere can deepen one's appreciation for the complexities of oenology and viticulture. As festivals continue to evolve and expand globally, the role of wine in these occasions will undoubtedly remain a cornerstone of cultural expression and enjoyment, further solidifying its place in the fabric of society.

The economic impact of wine festivals

Wine festivals have become a significant driver of economic activity in regions where viticulture is a predominant industry. These events attract both local residents and tourists, stimulating revenue for wineries, restaurants, hotels, and other businesses in the area. The economic impact of wine festivals extends beyond the immediate boost in sales during the event itself. They often result in increased awareness and interest in local wines, leading to long-term growth in wine tourism and sales. Additionally, these festivals create opportunities for networking and collaboration among industry professionals, fostering innovation and cooperation in oenology and viticulture. By showcasing the best wines of a region and providing a platform for industry experts to exchange ideas, wine festivals play a crucial role in driving economic growth and advancing the field of winemaking.

Wine's contribution to event experiences

Wine plays a significant role in enhancing event experiences, providing not only a sophisticated touch but also a sensory journey for participants. The selection of wine varietals, the art of wine pairing, and the presentation of wine tastings can elevate any event to a higher level of sophistication and enjoyment. Events such as wine tastings, wine pairings, and vineyard tours offer unique opportunities for attendees to engage with the culture and nuances of winemaking. By integrating wine into event experiences, organizers can create memorable and immersive experiences that cater to various preferences and interests. The intricate flavors, aromas, and textures of wine can stimulate the senses and create a multi-dimensional experience that adds depth and richness to any event. Ultimately, wine's contribution to event experiences showcases its versatility and ability to transform ordinary gatherings into extraordinary occasions.

Case studies of successful wine-centered events

Case studies of successful wine-centered events offer valuable insights into the effective implementation of innovative techniques and strategies in the realm of oenology and viticulture. By examining the success stories of events such as the Napa Valley Wine Auction and the Bordeaux Wine Festival, researchers can identify key factors that contribute to the popularity and profitability of these gatherings. These case studies provide a comprehensive understanding of the intricacies involved in organizing successful wine events, including the selection of top-quality wines, engaging guest experiences, and effective marketing and promotion strategies. Furthermore, they shed light on the importance of fostering collaborations between wineries, local businesses, and tourism boards to create vibrant and memorable experiences for attendees. Ultimately, these case studies serve as inspiration for future initiatives in the industry, highlighting the potential for continued growth and innovation in oenology and viticulture.

XLVI. THE INTERSECTION OF WINE AND TECHNOLOGY STARTUPS

In the ever-evolving landscape of oenology and viticulture, the intersection of wine and technology startups has emerged as a pivotal force driving innovation and progress in the industry. As traditional methods are being reevaluated and improved upon, these startups are revolutionizing the way wine is produced, marketed, and consumed. From AI-powered vineyard management systems to blockchain-based supply chain transparency, the integration of cutting-edge technologies is reshaping every aspect of the wine industry. Case studies showcasing the success of these startups underscore their potential to disrupt the status quo and create new opportunities for growth and sustainability. However, as these advancements continue to gain traction, it is crucial for researchers and practitioners to critically assess their implications and navigate the challenges that come with integrating technology into a centuries-old craft. This influx of innovation signals a new era for oenology and viticulture, emphasizing the need for continuous adaptation and forward-thinking strategies to ensure the industry's future success.

The emergence of wine tech startups

In the modern landscape of oenology and viticulture, the emergence of wine tech startups has become a defining feature of this revolutionary era. These innovative companies are pushing the boundaries of tradition by leveraging technology to enhance every aspect of the winemaking process, from vineyard management to fermentation control. Through the use of data analytics, artificial intelligence, and precision viticulture, these startups are revolutionizing the way wine is produced, marketed, and consumed. Case studies have demonstrated the significant impact of these advancements, showing improvements in quality, sustainability, and cost-effectiveness. As these startups continue to disrupt the industry, it is evident that the future of oenology and viticulture is being shaped by their ingenuity and vision. Looking ahead, the challenges and opportunities presented by these changes will undoubtedly drive further research and innovation in the field.

Disruptive technologies reshaping the wine industry

Disruptive technologies are fundamentally reshaping the wine industry, ushering in a new era of oenology and viticulture. The traditional methods of winemaking and grape cultivation are undergoing a profound transformation as advancements in digital technologies, precision viticulture, and sustainable practices take center stage. For instance, the adoption of sensor technology allows for real-time monitoring of vineyards, ensuring optimal conditions and resource management. Furthermore, artificial intelligence and machine learning algorithms are revolutionizing grape quality assessment and wine production processes, leading to enhanced efficiency and quality control. These disruptive technologies not only improve the overall productivity and sustainability of the industry but also open up new avenues for experimentation and innovation. As the wine industry continues to embrace these technological advancements, the future of oenology and viticulture appears to be filled with endless possibilities and exciting challenges.

The challenges and successes of wine tech entrepreneurship

In the realm of wine tech entrepreneurship, there are both formidable challenges and remarkable successes that shape the landscape of innovation in oenology and viticulture. The initial hurdles facing entrepreneurs in this field include navigating complex regulatory frameworks, securing funding for research and development, and gaining acceptance from traditional wine producers hesitant to embrace new technologies. However, those who successfully overcome these obstacles find themselves at the forefront of revolutionizing the industry. Through the utilization of cutting-edge technologies such as artificial intelligence, blockchain, and precision viticulture, wine tech entrepreneurs are improving overall efficiency in grape growing, winemaking processes, and marketing strategies. Successful ventures in this space have demonstrated the potential for increased sustainability, quality control, and consumer engagement. Despite the challenges that accompany wine tech entrepreneurship, the potential for transformative innovation in oenology and viticulture is immense, reshaping the future of the industry.

XLVII. WINE EDUCATION AND TECHNOLOGY TRANSFER

The integration of technology into wine education has been instrumental in revolutionizing the field of oenology and viticulture. By utilizing digital platforms, online courses, and virtual reality tools, wine education is now more accessible and engaging than ever before. This shift not only allows for a broader dissemination of knowledge but also fosters a deeper understanding of the intricate processes involved in winemaking. Additionally, technology transfer initiatives have propelled the implementation of cutting-edge techniques and methodologies across the industry, leading to improved wine quality and sustainability practices. Case studies showcasing the successful application of such advancements highlight the significant impact they have had on the field. As wine education continues to evolve in conjunction with technological innovations, the future holds great promise for further advancements in oenology and viticulture, shaping the landscape of the industry for years to come.

The importance of knowledge dissemination in viticulture

Knowledge dissemination plays a pivotal role in driving innovation and progress in viticulture. The sharing of information, research findings, and best practices among professionals, researchers, and enthusiasts in the field is essential for fostering growth and development. Through knowledge dissemination, new techniques, technologies, and methodologies can be introduced and implemented, leading to advancements in grape cultivation, wine production, and sustainability practices. Case studies and recent data showcase the impact that disseminating knowledge can have on the industry, from improving yield and quality of grapes to enhancing the overall efficiency of winemaking processes. As we navigate this new era of oenology and viticulture, the importance of sharing knowledge cannot be overstated. By actively engaging in knowledge dissemination, we can continue to push the boundaries of what is possible in the world of wine, shaping the future of the industry for generations to come.

Platforms and methods for technology transfer

One vital aspect of the new era in oenology and viticulture is the platforms and methods being utilized for technology transfer. In order to keep pace with the rapid advancements in the field, it is essential for researchers and practitioners to have access to cutting-edge technologies and innovative methodologies. Platforms such as online databases, virtual workshops, and collaborative research networks have emerged as crucial tools for sharing knowledge and facilitating the transfer of technology within the industry. By utilizing these platforms, researchers can access a wealth of information, collaborate with experts from around the world, and stay abreast of the latest developments in oenology and viticulture. These methods not only enhance the efficiency of technology transfer but also promote a culture of innovation and collaboration within the industry, ultimately steering it towards a more sustainable and prosperous future.

The role of academic institutions in wine education

Academic institutions play a crucial role in the realm of wine education, acting as the catalyst for innovation and progress within the field of oenology and viticulture. These institutions provide a platform for the dissemination of knowledge, research, and practical experience, shaping the minds of future winemakers and industry leaders. By offering specialized courses, workshops, and research opportunities, academic institutions equip students with the necessary skills and expertise to navigate the complexities of winemaking in a rapidly evolving industry. Moreover, these institutions serve as hubs for experimentation and collaboration, fostering a culture of continuous learning and discovery. Through partnerships with industry experts and research initiatives, academic institutions contribute to the development of cutting-edge techniques and technologies that are revolutionizing the way wine is produced, enhancing quality, sustainability, and innovation in the process. Ultimately, the role of academic institutions in wine education is instrumental in driving forward the new era of oenology and viticulture, shaping the future of the industry.

XLVIII. THE FUTURE OF WINE DISTRIBUTION CHANNELS

The landscape of wine distribution has undergone significant changes in recent years, driven by technological advancements and shifting consumer behaviors. Traditional distribution channels, such as retail stores and restaurants, are facing increasing competition from online platforms and direct-to-consumer sales. E-commerce has allowed wineries to reach a broader audience and establish a more direct connection with their customers, bypassing the need for intermediaries. Additionally, the rise of wine subscription services and wine clubs has provided consumers with curated selections tailored to their preferences. As we look to the future, it is clear that the traditional distribution model is evolving to meet the demands of a digital age. However, challenges such as regulatory restrictions and logistical complexities continue to shape how wine is distributed and consumed. As oenologists and viticulturists navigate this changing landscape, it will be crucial to adapt and innovate to ensure the industry's continued growth and success.

Trends in wine retail and distribution

They have seen significant changes in recent years, driven by advancements in technology and changing consumer preferences. Online sales and direct-to-consumer models have become increasingly popular, allowing wineries to reach a wider audience and bypass traditional distribution channels. Data analytics and AI tools are being used to personalize marketing strategies and improve the overall customer experience. Additionally, the rise of e-commerce platforms has made it easier for smaller producers to enter the market and connect with consumers on a global scale. However, these changes are also posing challenges, such as maintaining brand integrity and standing out in a crowded marketplace. As the industry continues to evolve, it is vital for wine retailers to adapt to these trends and embrace innovation in order to thrive in the ever-changing landscape of wine retail and distribution.

The impact of direct shipping and online sales

Direct shipping and online sales have significantly impacted the wine industry, revolutionizing the way consumers access and purchase their favorite wines. With the rise of e-commerce platforms and the ability for wineries to ship directly to consumers, traditional distribution channels are being disrupted. This shift allows small and medium-sized wineries to reach a broader audience, bypassing the need for distributors and retailers. By establishing a direct connection with consumers, wineries can better showcase their unique offerings and build brand loyalty. Additionally, online sales provide a convenient way for consumers to explore new wines and learn about different regions and varietals. With the increasing popularity of direct shipping and online sales, the wine industry is entering a new era of accessibility and innovation, reshaping the traditional landscape of oenology and viticulture.

The future landscape of wine distribution

The future landscape of wine distribution is undergoing a profound transformation in response to advancements in oenology and viticulture. With the rise of online platforms, direct-to-consumer sales, and the increasing popularity of natural and organic wines, the traditional models of wine distribution are being challenged. Wineries are now leveraging innovative technologies such as blockchain to authenticate and trace the origins of their products, ensuring transparency and quality for consumers. Additionally, the growing trend towards personalized subscription services and virtual tasting experiences is redefining how consumers interact with and purchase wine. As the industry continues to evolve, it is essential for wine distributors to adapt to these changes by embracing digital tools, enhancing customer experiences, and adopting sustainable practices to meet the demands of a dynamic marketplace. The future of wine distribution lies in harnessing technology, customer engagement, and sustainability to navigate the shifting landscape of the industry.

XLIX. THE ROLE OF WINE IN COMMUNITY BUILDING

In the evolution of community building, wine plays a crucial role in fostering social connections and creating a sense of belonging among individuals. Wine has a long-standing tradition of bringing people together, from ancient civilizations to modern societies. Through shared experiences of tasting, discussing, and appreciating wine, individuals can build relationships, strengthen bonds, and create a collective identity. Wine tastings, vineyard visits, and wine-related events provide opportunities for people to connect, communicate, and collaborate, enhancing social cohesion and building a sense of community. As a symbol of celebration, pleasure, and conviviality, wine transcends cultural boundaries and unites people across different backgrounds. Its ability to stimulate conversation, promote cultural exchange, and create shared memories makes wine a powerful tool for community building in today's interconnected world. In the era of revolutionary oenology, the role of wine in community building becomes even more significant, shaping the way we interact, connect, and engage with one another.

Wine clubs and their contribution to social cohesion

Wine clubs play a crucial role in fostering social cohesion within communities, bringing together individuals with a shared passion for oenology and viticulture. These clubs provide a platform for members to engage in meaningful discussions about wine, exchange knowledge, and cultivate lasting friendships. By organizing tastings, events, and educational programs, wine clubs create a sense of belonging and camaraderie among members. Additionally, these clubs often support local wineries and vineyards, thus contributing to the sustainability of smaller wine producers. Through shared experiences and mutual appreciation for wine, members of wine clubs strengthen social bonds and build networks that extend beyond the club itself. Ultimately, wine clubs serve as catalysts for social cohesion, enriching the lives of individuals and enhancing the overall wine culture within a community.

The impact of local wineries on community development

Local wineries play a crucial role in community development by contributing to economic growth, social cohesion, and cultural enrichment. These small-scale operations not only stimulate tourism and boost revenue for the local economy but also provide employment opportunities and support local businesses. By hosting events, tastings, and tours, wineries create a sense of place and identity, attracting visitors and residents alike. Furthermore, local wineries often source their grapes from nearby vineyards, thereby supporting local farmers and promoting sustainable agriculture practices. The relationships forged between winemakers, residents, and tourists foster a sense of community pride and engagement. As a result, the impact of local wineries on community development is multifaceted, creating a ripple effect that extends beyond the confines of the vineyard. In the era of revolutionary oenology, these local establishments are instrumental in shaping the future of the industry and fostering a sustainable and vibrant community ethos.

Wine as a medium for community engagement and support

Wine has transcended its traditional role as a beverage and is now becoming a powerful medium for community engagement and support. Through wine-related events, such as tastings, festivals, and fundraisers, individuals are coming together to celebrate and support causes they are passionate about. Wineries are also increasingly involved in community initiatives, partnering with local charities, businesses, and organizations to give back to their communities. This trend is not only fostering a sense of belonging and camaraderie among wine enthusiasts but also providing tangible benefits to those in need. By leveraging the social appeal of wine, communities are finding new ways to connect, support one another, and make a positive impact. This evolution in the role of wine signifies a shift towards a more inclusive and socially conscious approach to oenology, emphasizing the importance of community engagement in the world of wine.

L. THE ETHICS OF WINE PRODUCTION AND CONSUMPTION

The ethics of wine production and consumption in today's oenology and viticulture landscape are increasingly becoming a focal point for discussion and debate. As the industry continues to evolve, with new techniques and technologies being developed and implemented, the ethical considerations surrounding the environmental impact, labor practices, and consumer health implications are more critical than ever. From sustainable farming practices to fair labor initiatives, there is a growing recognition of the need for ethical standards to be upheld throughout the entire wine production chain. Consumers are also becoming more conscious of the ethical implications of their wine choices, demanding transparency and accountability from wineries. As we navigate this new era of oenology, it is essential to consider not only the quality of the wine but also the ethical principles that guide its production and consumption, ultimately shaping the future of the industry.

Ethical considerations in labor practices and sourcing

A crucial aspect of the revolutionary era in oenology and viticulture is the ethical considerations in labor practices and sourcing. Labor practices in vineyards and wineries have faced scrutiny regarding fair wages, working conditions, and treatment of workers. Ethical sourcing involves the procurement of grapes and other materials in a way that respects the environment, supports local communities, and ensures transparency in the supply chain. It is essential for the wine industry to address these issues to maintain sustainability and credibility in the eyes of the consumer. By implementing ethical labor practices and sourcing guidelines, wineries and vineyards can not only enhance their reputation but also contribute positively to social and environmental concerns. The incorporation of ethical considerations in labor practices and sourcing will play a significant role in shaping the future of oenology and viticulture, emphasizing a holistic approach towards sustainability and responsible business practices.

The social responsibilities of wine consumers

As key players in the wine industry, consumers have a duty to consider the environmental and social impact of their purchasing decisions. By choosing wines produced sustainably and ethically, consumers can support vineyards and wineries that prioritize responsible practices. This includes factors such as organic farming methods, fair labor practices, and biodiversity conservation. Furthermore, consumers can advocate for transparency in labeling and marketing to ensure that they are making informed choices. Embracing these social responsibilities can not only contribute to a more sustainable and ethical wine industry but also empower consumers to drive positive change in the sector. Therefore, wine consumers play a crucial role in promoting a more environmentally conscious and socially responsible wine industry.

The role of ethics in shaping the future of wine

Ethics play a pivotal role in shaping the future of wine, as consumers are increasingly demanding transparency and sustainability in the production process. Winemakers are now faced with the challenge of integrating ethical practices into every aspect of their operations, from vineyard management to labeling and marketing. For example, the rise of organic and biodynamic viticulture reflects a growing awareness of environmental concerns and a desire to minimize chemical inputs. Additionally, fair labor practices and social responsibility are becoming increasingly important considerations in the wine industry, as consumers seek to support ethical producers. By embracing ethical principles, winemakers can not only meet consumer demand but also contribute to a more sustainable and socially responsible future for the industry. As such, ethics will continue to be a driving force in shaping the direction of oenology and viticulture in the years to come.

LI. CONCLUSION

In conclusion, the field of oenology and viticulture is experiencing a profound shift towards innovation and advancement. The rise of new techniques, technologies, and methodologies has opened up a world of possibilities for winemakers and grape growers alike. Through case studies and recent data, it is evident that these developments are not only improving the quality of wine but also revolutionizing the entire industry. As we look to the future, it is essential to consider how these changes will continue to shape the landscape of oenology and viticulture. Challenges such as climate change, sustainability, and consumer preferences will need to be addressed through continued research and practice. The new era of oenology and viticulture is upon us, promising exciting opportunities and potential for growth in the years to come.

Summary of the revolutionary changes in oenology and viticulture

The revolutionary changes in oenology and viticulture have brought about a seismic shift in the way wine is produced and consumed. Innovations in techniques, technologies, and methodologies have played a significant role in this transformation. For instance, the introduction of precision viticulture has enabled vineyard managers to make data-driven decisions, leading to improved grape quality and yield. Additionally, advancements in fermentation processes, such as the use of wild yeast strains, have allowed winemakers to create more complex and diverse flavors in their wines. The adoption of sustainable practices, like organic and biodynamic farming, has also gained traction in the industry, reflecting a growing concern for environmental sustainability. These changes are not only reshaping the current landscape of oenology and viticulture but are also setting the stage for a future that is more innovative and forward-thinking. As we look ahead, it is crucial for researchers and practitioners to continue exploring new possibilities and challenges that arise in this dynamic field.

Reflection on the implications for the wine industry's future

Reflection on the implications for the wine industry's future unveils a landscape ripe with possibilities and challenges. The revolutionary advancements in oenology and viticulture are set to redefine the way wine is produced, perceived, and consumed. The integration of cutting-edge technologies, such as precision viticulture and artificial intelligence, promises increased efficiency, quality, and sustainability in grape growing and winemaking processes. Moreover, the adoption of sustainable practices and organic farming methods reflects a growing consumer demand for environmentally conscious products. However, as the industry evolves, it faces potential hurdles such as climate change, economic pressures, and shifting consumer preferences. The future of the wine industry will require a delicate balance between tradition and innovation, where stakeholders must adapt to emerging trends while preserving the heritage and artistry of winemaking. It is essential for researchers and practitioners to anticipate these challenges and opportunities, paving the way for a dynamic and prosperous future in oenology and viticulture.

Challenges and opportunities for ongoing research and innovation

Challenges and opportunities for ongoing research and innovation in oenology and viticulture are abundant in the current revolutionary era. One significant challenge is the need to adapt to changing climate conditions, which can affect grape yields and quality. Researchers are exploring new techniques such as precision viticulture and climate-resilient grape varieties to mitigate these challenges. Another opportunity lies in the integration of data-driven technologies such as artificial intelligence and big data analytics into winemaking processes. These innovations can enhance quality control, optimize production efficiency, and improve decision-making in vineyard management. However, the adoption of these technologies may pose challenges related to data privacy and security. As ongoing research delves further into these areas, it is crucial to address these challenges and seize the opportunities presented by technological advancements to shape the future of oenology and viticulture positively.

BIBLIOGRAPHY

Mariola Staniak. 'Sensory Nudges.' The Influences of Environmental Contexts on Consumers' Sensory Perception, Emotional Responses, and Behaviors toward Food and Beverages, Han-Seok Seo, MDPI, 9/9/2021

Alan Young. 'Making Sense of Wine.' A Study in Sensory Perception, Greenhouse Publications, 1/1/1986

Howard Moskowitz. 'Consumer-based New Product Development for the Food Industry.' Sebastiano Porretta, Royal Society of Chemistry, 4/7/2021

Charles G. Edwards. 'Wine Microbiology.' Practical Applications and Procedures, Kenneth C. Fugelsang, Springer Science & Business Media, 4/3/2007

John F. Jackson. 'Wine Analysis.' Hans-Ferdinand Linskens, Springer Science & Business Media, 12/6/2012

Gavin L. Sacks. 'Understanding Wine Chemistry.' Andrew L. Waterhouse, John Wiley & Sons, 6/6/2016

Riccardo Flamini. 'Hyphenated Techniques in Grape and Wine Chemistry.' John Wiley & Sons, 4/30/2008

Andrew G. Reynolds. 'Managing Wine Quality.' Volume 2: Oenology and Wine Quality, Woodhead Publishing, 12/3/2021

Antonio Morata. 'Red Wine Technology.' Academic Press, 10/29/2018

Alexander Hollaender. 'Genetic Control of Environmental Pollutants.' Gilbert S. Omenn, Springer Science & Business Media, 11/11/2013

Graham H. Fleet. 'Yeasts in Food and Beverages.' Springer Science & Business Media, 1/10/2006

Joshua Rosenthal. *'Biodiversity and Human Health.'* Francesca Grifo, Island Press, 2/1/1997

Robert B. McKinstry. *'Biodiversity Conservation Handbook.'* State, Local, and Private Protection of Biological Diversity, Environmental Law Institute, 1/1/2006

Jutta Stadler. *'Biodiversity and Health in the Face of Climate Change.'* Melissa R. Marselle, Springer, 6/11/2019

Rebekka Schütte. *'Supporting Biodiversity in European Vineyards: Possibilities for Winegrowers and Economic Implications.'* Georg-August-Universität Göttingen, 1/1/2019

William J. Slattery. *'Tabulation of Voluntary Standards and Certification Programs for Consumer Products.'* Department of Commerce, National Bureau of Standards, Institute for Applied Technology, Standards Application and Analysis Division, 1/1/1977

Willy Schilthuis. *'Biodynamic Agriculture.'* Floris, 1/1/2003

Luann Preston-Wilsey. *'Toward a Sustainable Wine Industry.'* Green Enology Research, CRC Press, 5/6/2015

Sofia Catarino. *'Improving Sustainable Viticulture and Winemaking Practices.'* J. Miguel Costa, Academic Press, 3/19/2022

Maria Manuela Chaves. *'Grapevine in a Changing Environment.'* A Molecular and Ecophysiological Perspective, Hernâni Gerós, John Wiley & Sons, 10/5/2015

Herman Casteleyn. *'Conserve Water, Drink Wine.'* Recollections of a Vinous Voyage of Discovery, Ronald S Jackson, Taylor & Francis, 1/10/1997

K. William Easter. *'Irrigation Investment, Technology, And Management Strategies For Development.'* Routledge, 4/3/2019

Ted Goldammer. *'The Grape Grower's Handbook.'* A Guide to Viticulture for Wine Production, Apex Publishers, 1/1/2018

Ram C Dalal. *'Sustainable Soil Management.'* Beyond Food Production, Somasundaram Jayaraman, Cambridge Scholars Publishing, 6/15/2023

A. L. Page. 'Methods of Soil Analysis, Part 3.' Chemical Methods, D. L. Sparks, John Wiley & Sons, 1/22/2020

Mark Krstic. 'Healthy Soils for Healthy Vines.' Soil Management for Productive Vineyards, Robert White, Csiro Publishing, 9/1/2019

Kenroy C. Wedderburn. 'Compulsive Buying.' Consumer Traits, Self-Regulation, and Marketing Ethics, Trevor A. Smith, Rowman & Littlefield, 11/3/2021

David Hemming. 'Plant Sciences Reviews 2010.' CABI, 1/1/2011

Jose-Miguel Martinez-Zapater. 'Genetics, Genomics, and Breeding of Grapes.' Anne-Francoise Adam-Blondon, CRC Press, 4/19/2016

Vandana Mangal. 'Business And Information Technologies (Bit) Project, The: A Global Study Of Business Practice.' Uday S Karmarkar, World Scientific, 2/15/2007

Tony Proffitt. 'Precision Viticulture.' A New Era in Vineyard Management and Wine Production, Winetitles, 1/1/2006

Francesca Orlando. 'Impact of Agricultural Practices on Biodiversity of Soil Invertebrates.' Stefano Bocchi, MDPI, 1/6/2021

Chris Winefield. 'Resilience of Grapevine to Climate Change: From Plant Physiology to Adaptation Strategies.' Chiara Pastore, Frontiers Media SA, 9/20/2022

Franziska Hübsch. 'The impact of climate change on viticulture in Poland and Portugal.' GRIN Verlag, 4/15/2021

John Gladstones. 'Wine, Terroir and Climate Change.' Wakefield Press, 1/1/2011

Edward Cressy. 'Discoveries and Inventions of the Twentieth Century.' E.P. Dutton, 1/1/1915

John B. Cella. 'The Cella Family in the California Wine Industry.' Regional Oral History Office, Bancroft Library, University of California, 1/1/1986

Daniel Pambianchi. 'Techniques in Home Winemaking.' A Practical Guide to Making Château-Style Wines, Vehicule Press, 8/1/2011

Stuart J. Fleming. 'The Origins and Ancient History of Wine.' Food and Nutrition in History and Antropology, Patrick E. McGovern, Routledge, 9/2/2003

H. Panda. 'The Complete Book on Wine Production.' How to start a wine business, How to Start a Wine Production Business, How to Start a Wine Production industry?, How to start a wine store, How to start wine factory in India, How to Start Wine Making Business, How to Start Wine Production Industry in India, How to Start Winery Business, How to start your own wine business, How to Start Your Own Winery Business, How White Wines Are Made, How wine is made, Indian Wine Industry, Making Wine from Grapes, Making Wine, Most Profitable Wine Production Business Ideas, New small scale ideas in Wine Production industry, Niir Project Consultancy Services, 10/2/2011

Jean L. Jacobson. 'Introduction to Wine Laboratory Practices and Procedures.' Springer Science & Business Media, 6/14/2006